IMAGES
of America

Tennessee's Great Copper Basin

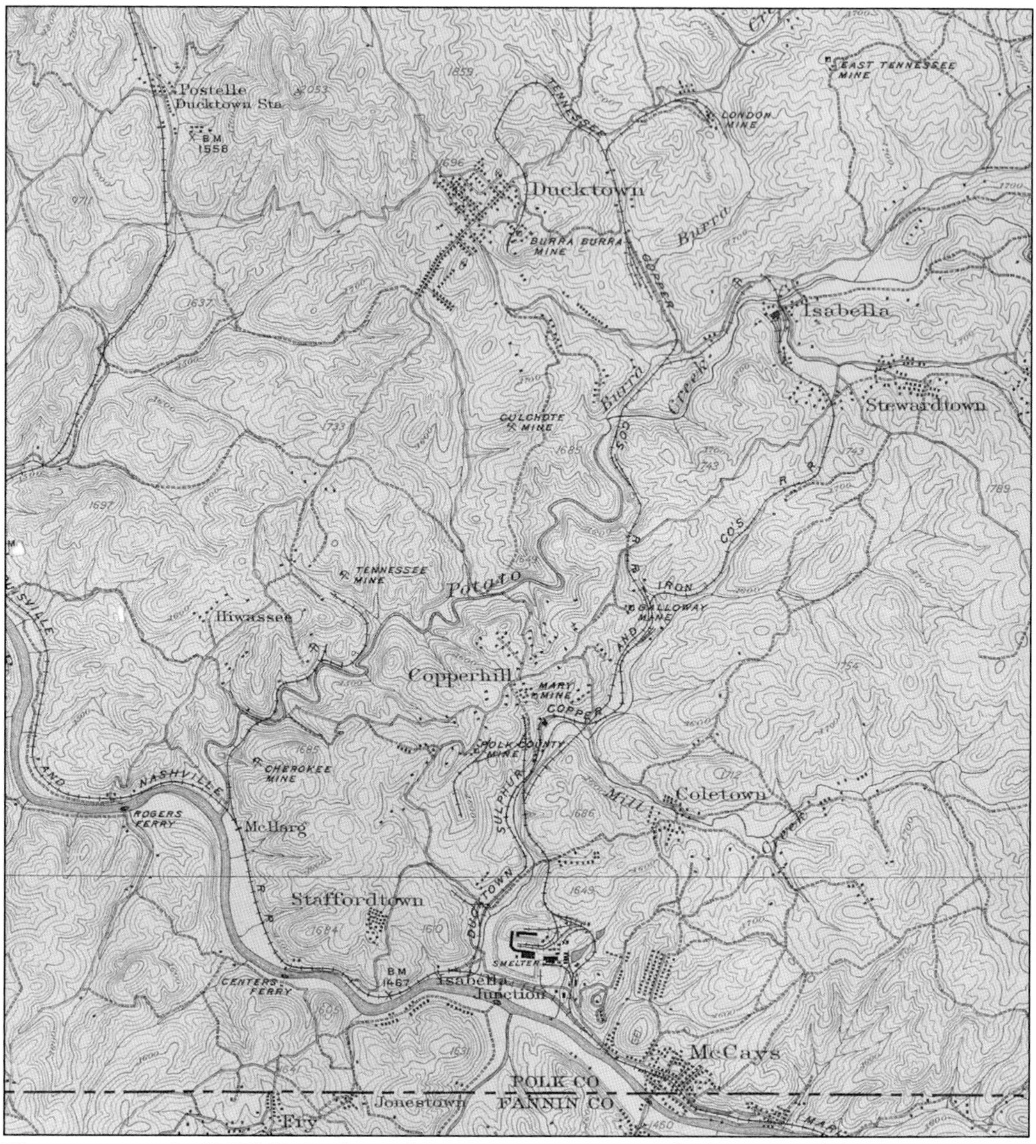

This segment of a 1907 US Geological Survey map pinpoints the location of the existing mines and communities in the Ducktown district at that time. It also shows the original site of the Copperhill Post Office, which was located at Mary Mine. The town that was soon to become Copperhill was still known as McCays, Tennessee. (Courtesy of the US Geological Survey.)

On the Cover: This photograph of employees on the settler floor at the Tennessee Copper Company (TCC) smelter was taken in 1904 by a young engineer named E.H. Westlake, who would later become corporate president of TCC. A number of Westlake's photographs from his tenure in the Ducktown Basin between 1903 and 1906 are being published for the first time in this book. (Courtesy of the Ducktown Basin Museum.)

IMAGES
of America

Tennessee's Great Copper Basin

Harriet Frye

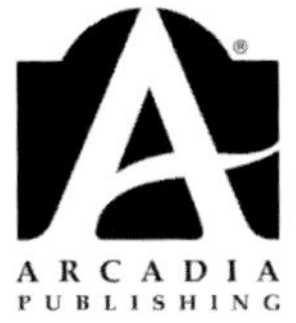

ISBN 978-1-4671-2494-2

Published by Arcadia Publishing
Charleston, South Carolina

Printed in the United States of America

Library of Congress Control Number: 2016949798

For all general information, please contact Arcadia Publishing:
Telephone 843-853-2070
Fax 843-853-0044
E-mail sales@arcadiapublishing.com
For customer service and orders:
Toll-Free 1-888-313-2665

Visit us on the Internet at www.arcadiapublishing.com

Dedicated to my parents, Robert Lowery Frye and Gertie Forrester Frye, without whom neither I nor this book would ever have come into being.

Contents

ACKNOWLEDGMENTS

A book like this has many authors, not just one. In fact, the first person I need to thank is the late R.E. Barclay, chief clerk in the Tennessee Copper Company General Office, whose three meticulously researched volumes of local history are still the definitive source for those of us intent on helping to preserve that history for future generations. Thank you, Mr. Barclay; little did you know, on that day in 1943 when you wrote in your daily log, "Frye out; his wife gave birth to a baby girl today," that our paths as authors would cross again more than seven decades later, right here in Copperhill.

Second, but by no means second in importance, I offer my particular thanks to the Ducktown Basin Museum; to curator Ken Rush, who patiently answered my questions, supplied me with detailed background material, and reviewed the final text for accuracy; to guides Joyce Allen and Dawna Standridge, who frequently and cheerfully allowed me to stay until the last possible minute just so I could scan one more photograph; and to all the members and friends, including the late Ron Henry and his widow, Loretta, who have donated images to the museum over the years. Unless otherwise noted, the images in this book, many of which are being printed here for the first time, are from the collection of the museum.

The members of two active and enthusiastic local-history Facebook groups have been indispensable in helping me solve a few photographic mysteries. If I had a question about a particular image, somebody in one of those groups very often had the answer. I also thank the many people who generously shared their personal and family photographs with me; I only wish I could have used them all.

Last, and certainly not least, I must recognize my Arcadia editor, Liz Gurley, not only for her general guidance, but also for stepping up to the plate and relieving me of some tedious detail work when I dislocated my elbow two weeks before my deadline. Thanks, Liz. I truly could not have done this without you.

Introduction

The mapmakers called it "Potato Creek." The locals called it "Tater." On a morning in August 1843, when the early sun revealed that the red crystals he had found there the evening before were not gold but merely copper, a disappointed prospector named Lemmons would probably have been justified in calling it something less printable.

At this point, the unfortunate Lemmons fades from the pages of local history. Yet, without him, that history might not have been written at all.

Seven years of exploration followed—tentative at first, but increasingly more productive as time went on. Finally, in May 1850, the first two shafts were sunk near the village of Hiwassee (the site of the present-day city of Ducktown), and copper was found just 12 feet below the surface.

By 1852, investors from as far away as London had descended on the Ducktown Basin, as the region was already known to geologists, with capital in hand. By the end of 1854, no less than 28 mining companies had been incorporated. The most significant of these would prove to be the New York–based Union Consolidated Mining Company, which purchased a group of local mining properties in 1858 and named German mine captain J.E. Raht as superintendent. Two years later, in a surprisingly noncompetitive spirit, the newly incorporated Burra Burra Copper Company also made Raht the superintendent of its operations at Ducktown.

Then came the Civil War. By 1863, with Cleveland, Tennessee, in the hands of Union forces and the local rail lines destroyed so that the Confederacy was unable to bring supplies into the area or ship the copper out, the mines were closed. The postwar years brought a partial return to the region's prewar prosperity. By 1878, however, even the venerable Union Consolidated Mining Company had found itself financially unable to go on.

The second period of prosperity, which began with the purchase of the old Union Consolidated properties in 1889 by the London-based Ducktown Copper, Sulphur & Iron Company Ltd. and continued with the formation of the Tennessee Copper Company in 1899, would prove to be of longer duration. This period, continuing until the early 1960s, is the primary focus of this book.

In July 1987, after three decades of gradual attrition, the last copper mines of the Copper Basin were permanently closed and allowed to flood. In December 2000, the last sulfuric acid plant at Copperhill ceased production. In 2017, as this book goes to press, the largest environmental cleanup and reclamation project in Tennessee history is nearing its end, and the region now known to most of its residents as the Copper Basin—or, more commonly, "the Basin"—has evolved into a world-class whitewater venue and one of Tennessee's most popular destinations for hiking and mountain biking. The waters of the Ocoee River run deep and clear again, the mountain air is fresh and pure, and the haunting red hills remembered by older residents are nowhere to be seen.

The creek, by the way, is now officially called "North Potato."

On this February 1876 pay sheet from the Isabella Smelter, monthly wages range from $6.50 for office boy James R. Jeffrey to $65 for blacksmith H.P. Kimsey. The monthly pay rate for matte furnace helpers was $32.50, the rate for yard laborers was $22.10, and a coal roller earned just $13 a month. Most of the workers drew considerably less, due to deductions for their accounts at the company store.

One

The Golden Age of Copper

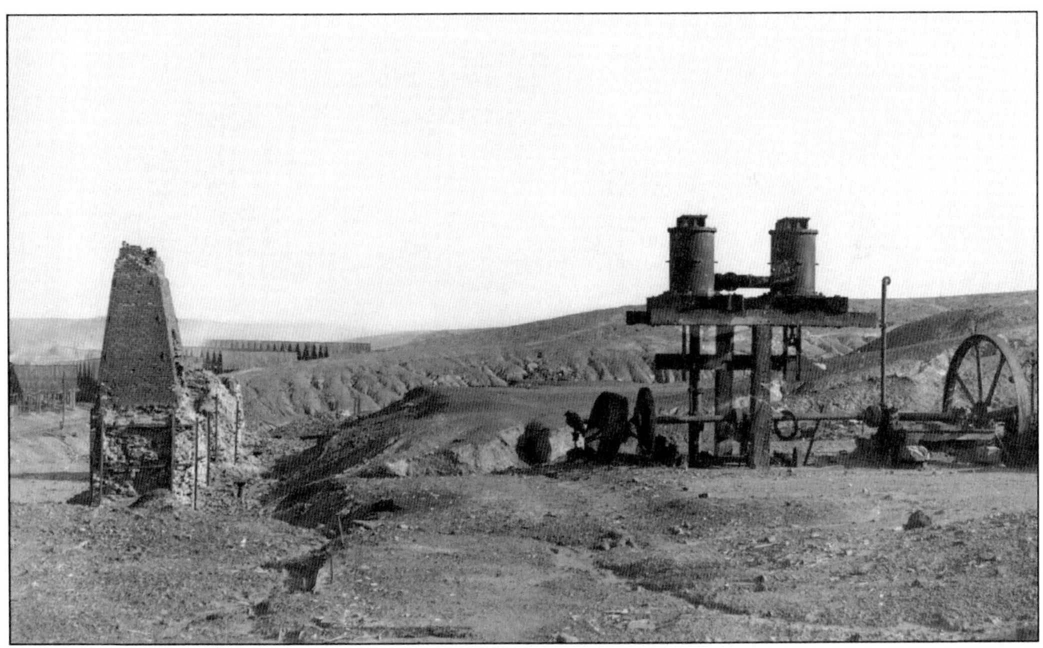

The ruins of the old Burra Burra smelter, erected around 1860, were photographed in 1904 by E.H. Westlake. The smelters of this era were powered first by water wheels and later by steam engines. In the decade immediately following the Civil War, Burra Burra was one of the most productive mines in the Basin, but the original mine and smelter were officially closed in 1877 due to mounting debt.

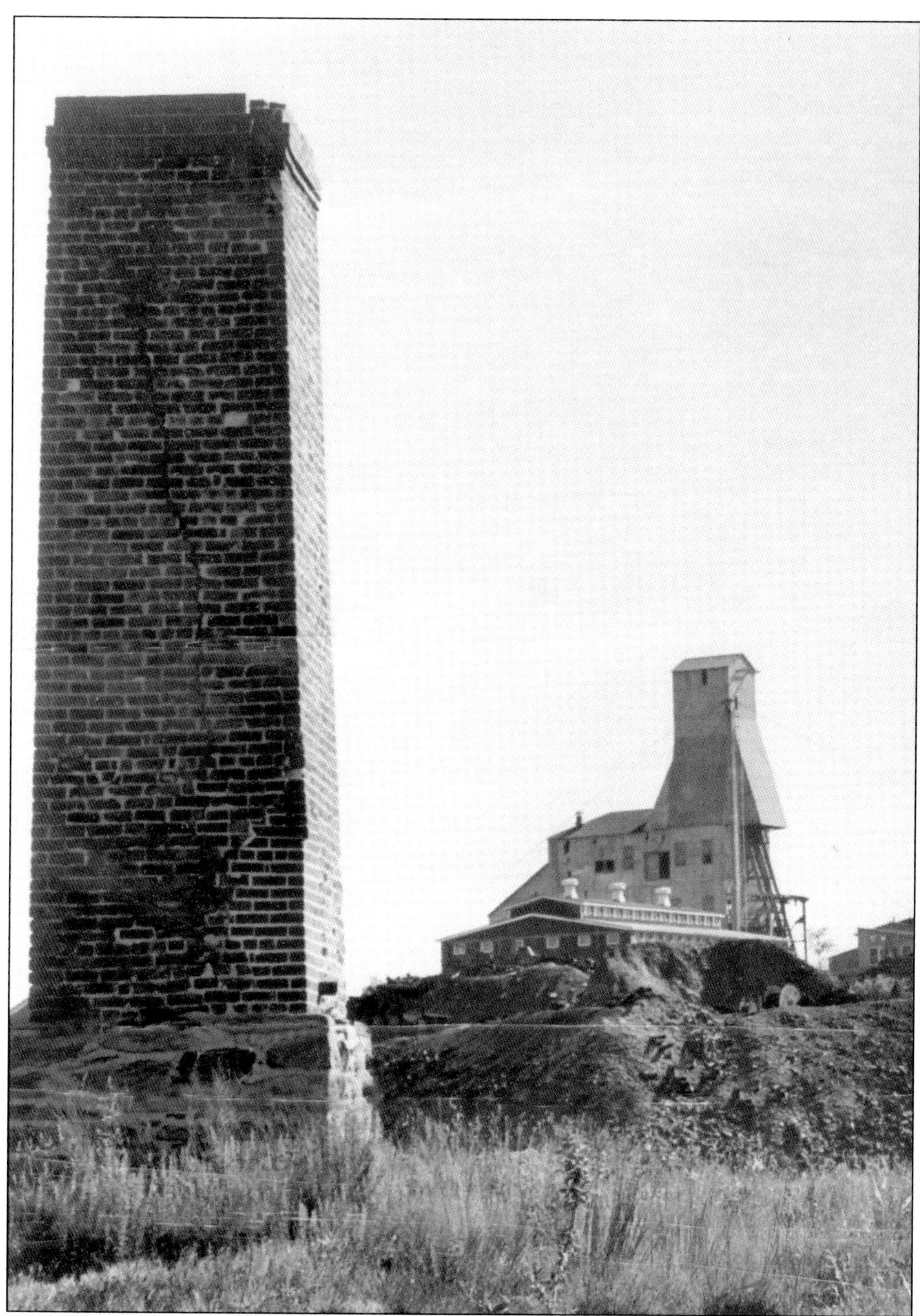

The original brick chimney of Hiwassee Mine and the headframe of Burra Burra Mine rise from separate hilltops overlooking the city of Ducktown. Hiwassee Mine, which was opened in 1850, was the earliest mine to be worked in the Basin. Although mining at Burra Burra had begun around 1860, the mine marked by the headframe in this photograph was opened by the Tennessee Copper Company in 1899 in a different location. The headframe was dismantled after the closing of Burra Burra in 1958, but the reconstructed Hiwassee chimney still stands in its original site on the outskirts of Ducktown.

During the early days of mining in the Ducktown district, experienced Cornish miners were in great demand. Mine captain John Quintrell, who arrived in 1854, was among the first wave of emigrants from Cambourne, Cornwall. This photograph of Quintrell and his second wife, Sara Jory, was taken around 1900. Sara was the daughter of another early Cornish immigrant, Mary Mine captain James Jory. (Courtesy of Grace Quintrell Postelle.)

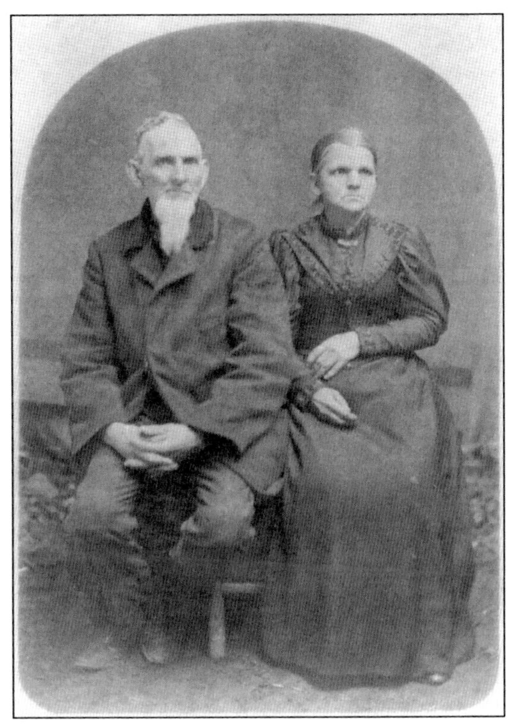

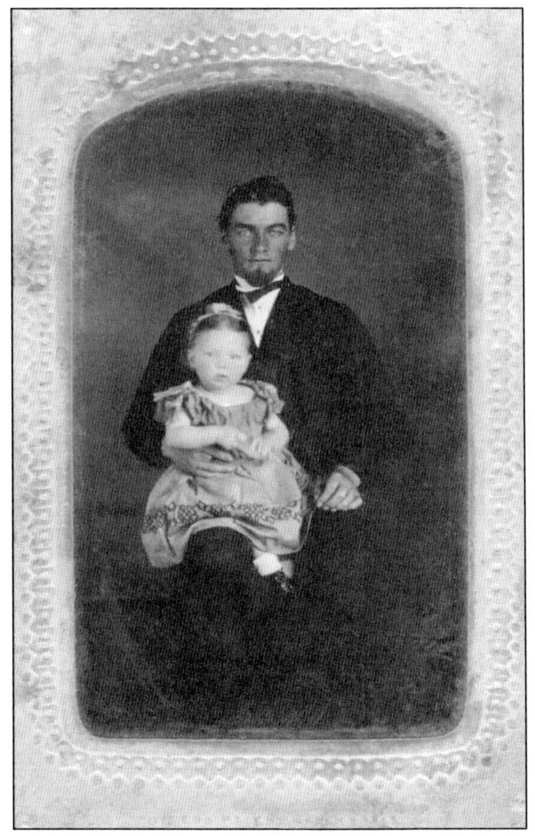

Cornish miner John Vellenoweth poses with his young daughter Elizabeth in this early 1870s tintype, which was taken after the family's arrival in the Ducktown district. Vellenoweth's wife, Mary Jory, was the daughter of Mary Mine captain James Jory and the sister of John Quintrell's second wife, Sara Jory. (Courtesy of Brian Smith.)

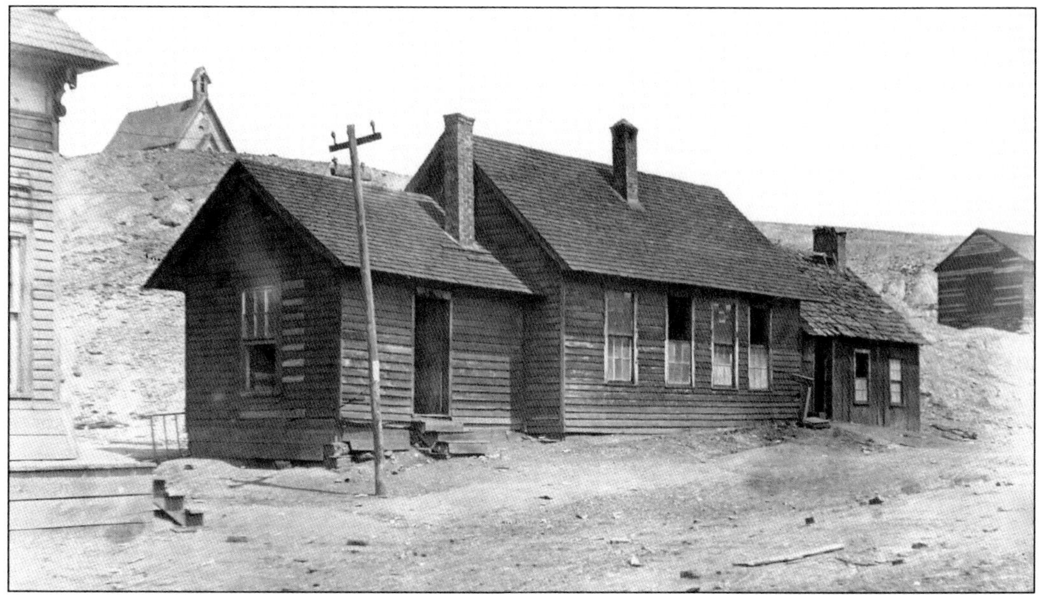

The old laboratory building at Ducktown Sulphur, Copper & Iron Company (DSC&I), constructed in the 1860s, was still in use when chemist Thorn Smith took this photograph in 1900. "It was equipped mainly with French labeled bottles," he wrote later, "but I knew what the labels meant. The building was heated by an old-fashioned stove with a lid in the top which we used to cook down our solutions."

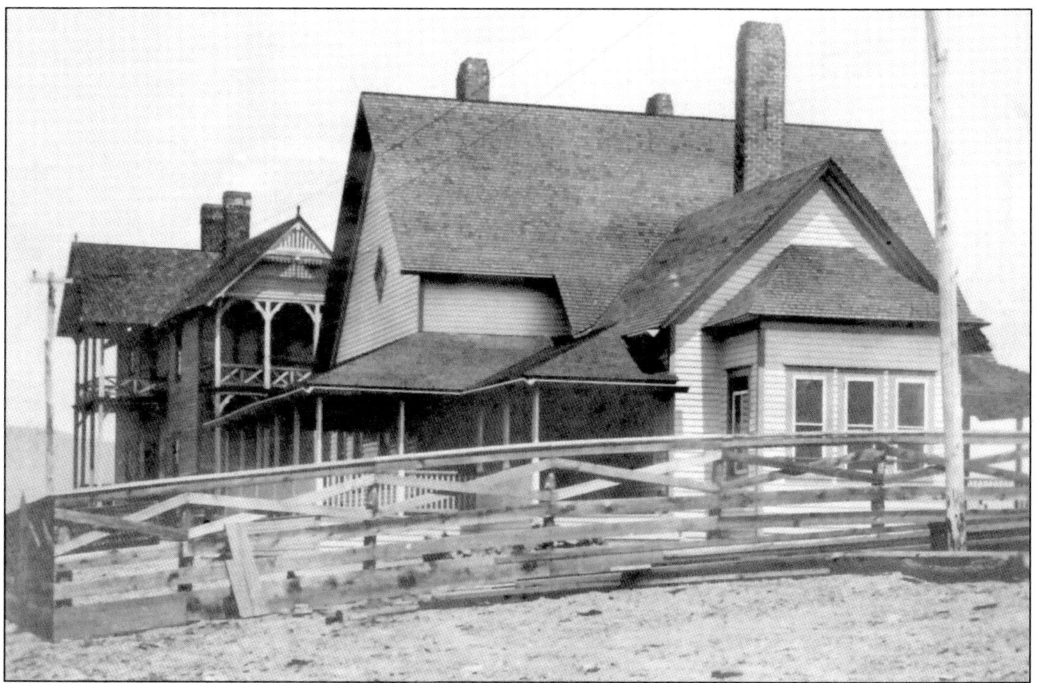

The first DSC&I staff house in Isabella sat on a hilltop about a mile from the mine works. According to Thorn Smith, who took this photograph around 1900, "All the minor employees lived there, the surveyor, the assistant superintendent and the like." Smith and his family lived here briefly before moving into company housing.

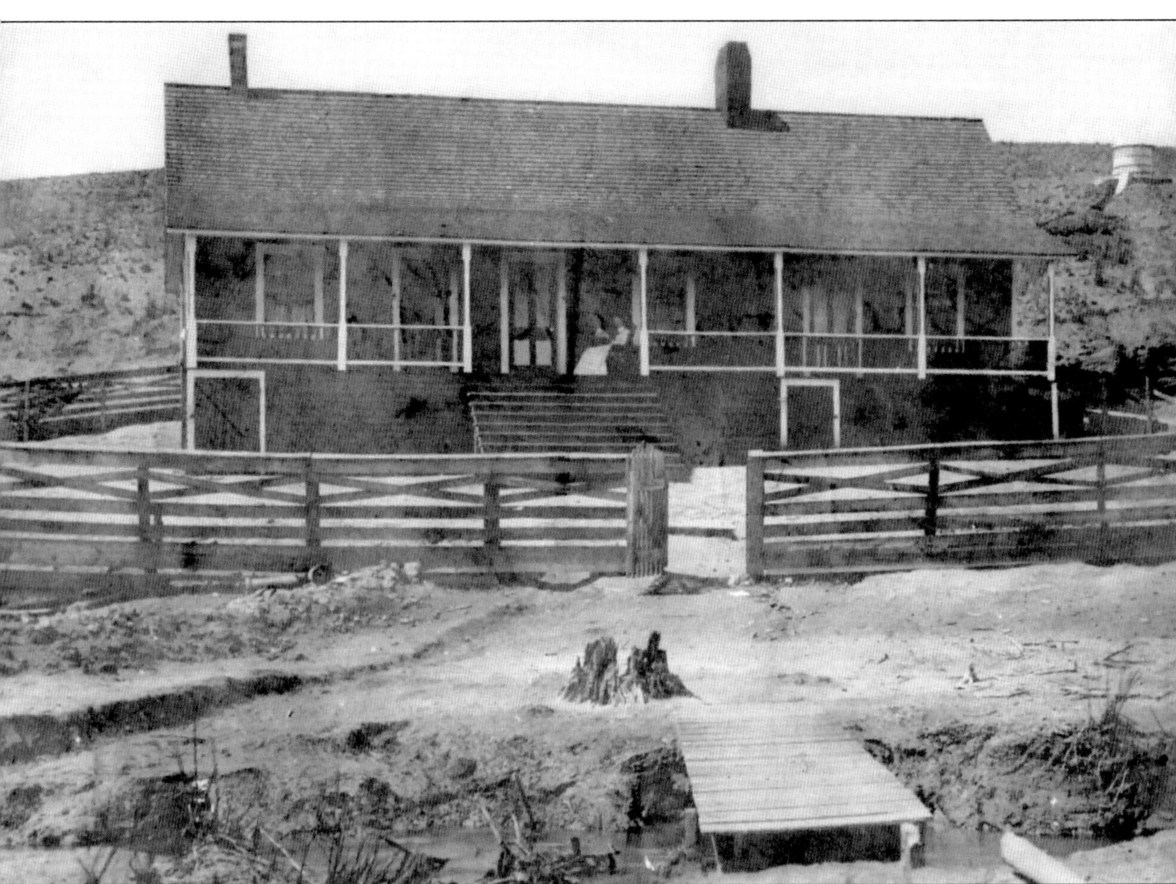

When Thorn Smith and his family moved into their second company house in Isabella in 1901, the luxury of their new quarters came as a great relief. Of the more primitive house they had initially rented, he wrote later, "Very soon . . . we discovered that every crack, and there were thousands of them, were the homes, one next to the other, of what one calls bedbugs. We fumigated with sulphur but they had been brought up on sulphur dioxide and thrived on it." Both houses rented for the same $5 out of Smith's $125 monthly salary, but the new house came without the bedbugs. "There was a high board fence around the place," he wrote, "and peace inside." On most days, he recalled, the Smith children could play safely in the little stream outside the fence. During heavy rains, however, the stream was subject to flash flooding. "[It] became a boiling mass of muddy water a hundred feet wide and ten feet deep," he wrote. "Sometimes it entered the yard and came up to the high front steps."

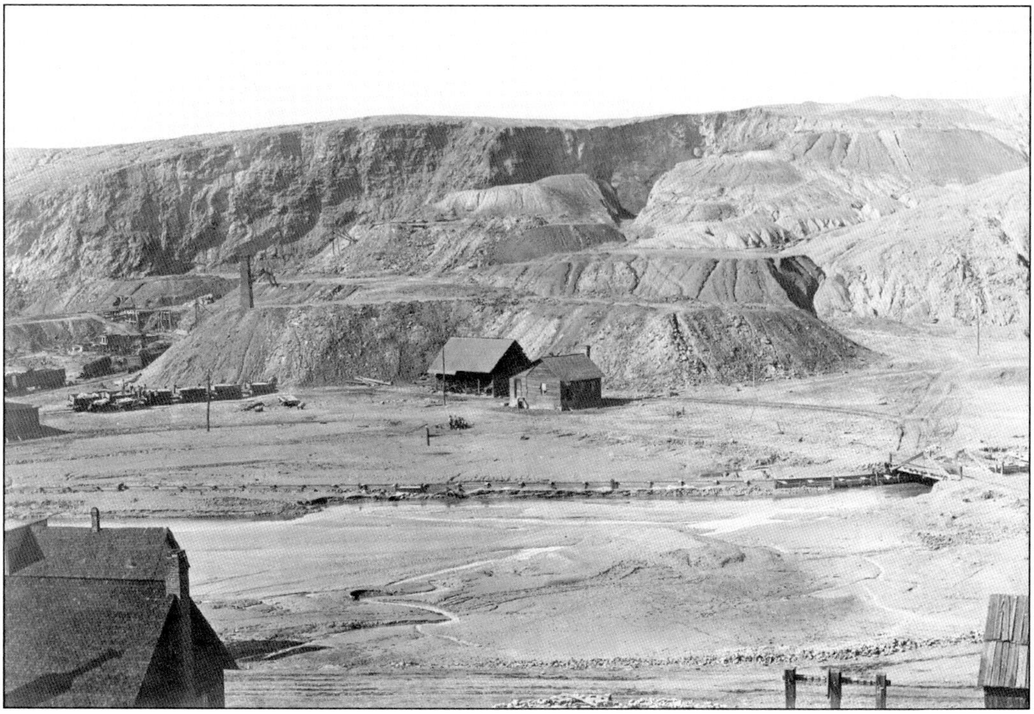

In 1889, when DSC&I set up its new headquarters at Isabella, sulfur fumes from years of open smelting had already turned the surrounding hills into an eroded landscape of red clay. In this early 1900s photograph, the chimney and part of the mine workings can be seen at left. The old company store and the rest of the small Isabella business district would have been off camera to the left.

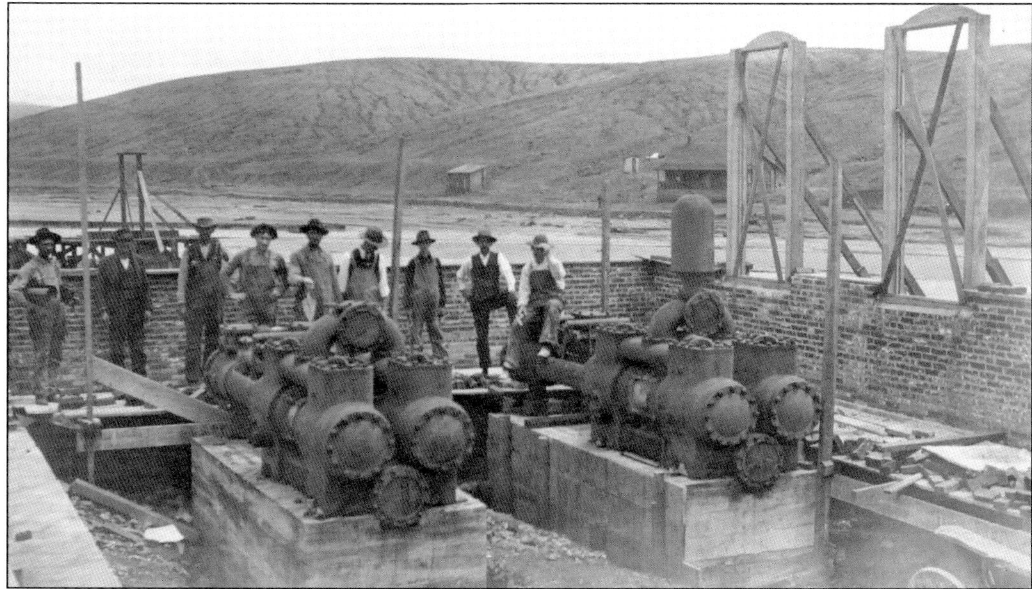

The pump house/power house at Isabella is still under construction in this 1890s photograph. By the time DSC&I took over the Isabella properties, the plant had been idle for more than a decade, and considerable work was needed to bring the facilities up to date.

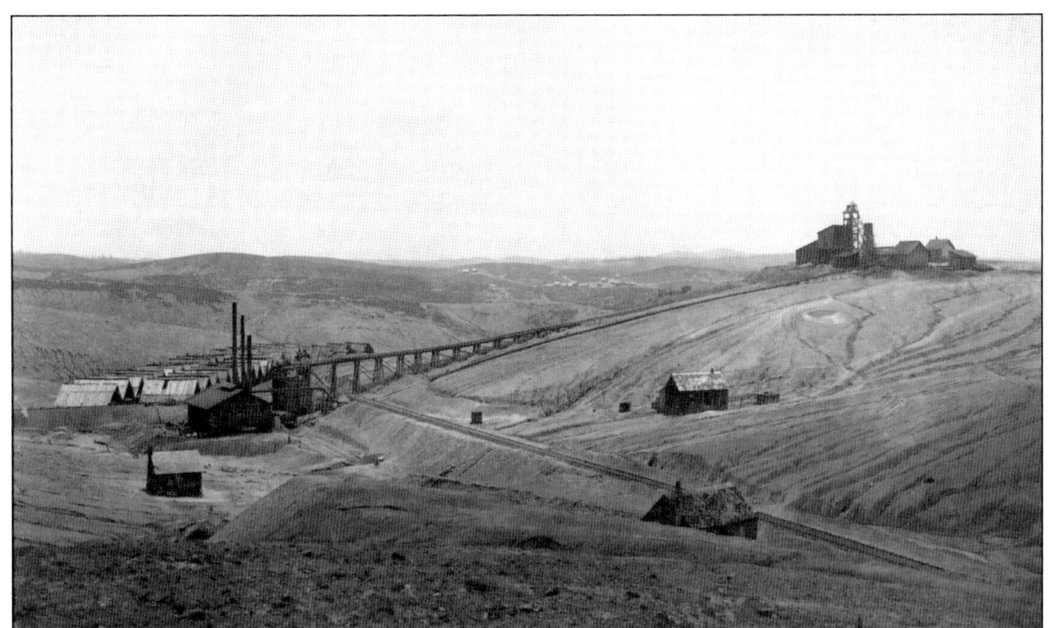

Calloway Mine, first opened in 1853, was reopened after its acquisition by DSC&I and produced 77,409 tons of ore between 1897 and 1907. The mine was located about a half-mile to the northeast of Mary Mine. In 1907, it was closed and allowed to flood. It was reopened in 1943 by TCC and remained in operation until 1983. This photograph was taken in 1904 by E.H. Westlake.

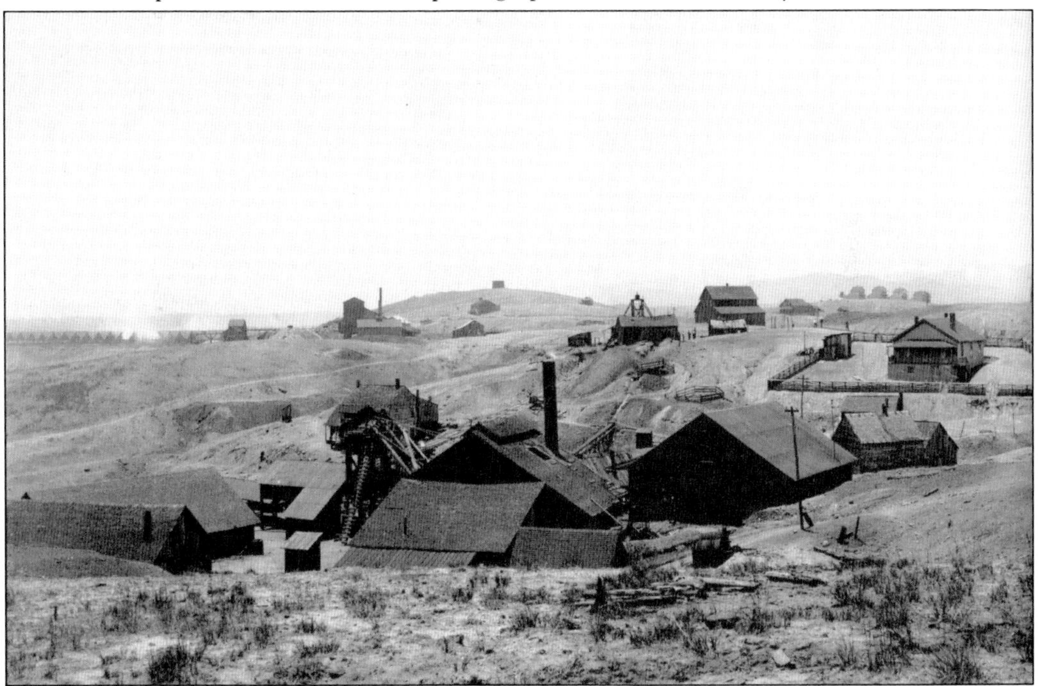

E.H. Westlake took this photograph in 1904, with DSC&I's Mary Mine in the foreground and TCC's Polk County Mine in the background. The two mines were located within sight of each other on the same vein of ore. After TCC acquired the DSC&I properties in 1936, the vein was mined as a single ore body, with operations based at Mary Mine.

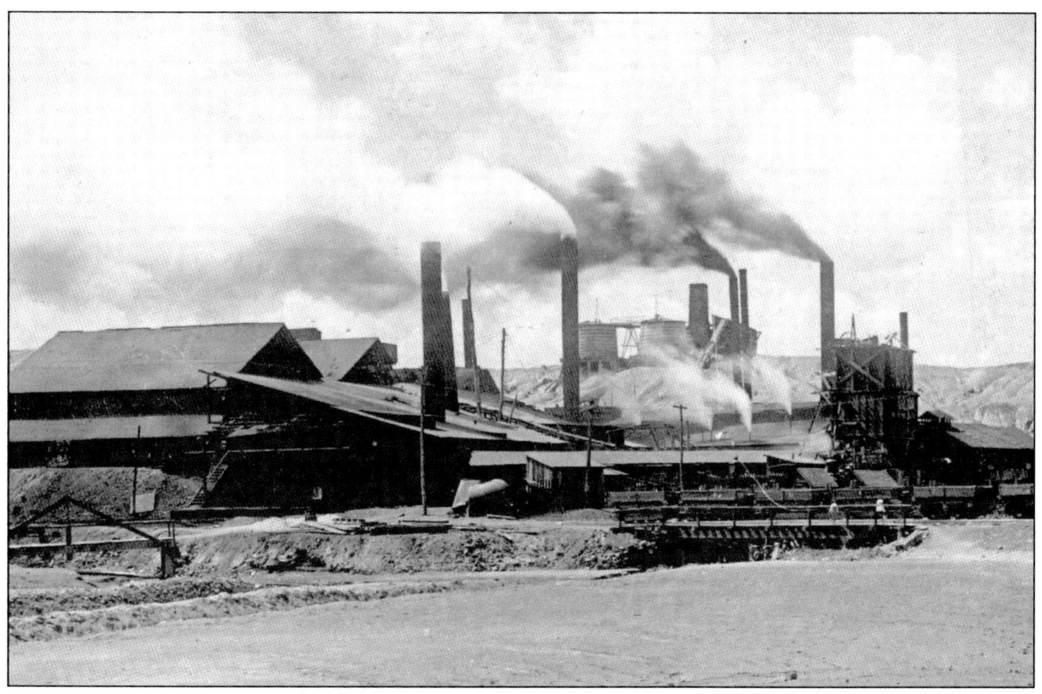

In 1903, DSC&I became the first company in the northern hemisphere to convert to pyritic smelting. As this 1904 photograph by E.H. Westlake makes clear, however, the changeover did little to eliminate the toxic sulfur fumes that had already embroiled the company in a series of smoke-damage lawsuits. In May 1907, in a decision written by Oliver Wendell Holmes, the US Supreme Court ruled against both DSC&I and Tennessee Copper Company for smoke damage in five Georgia counties.

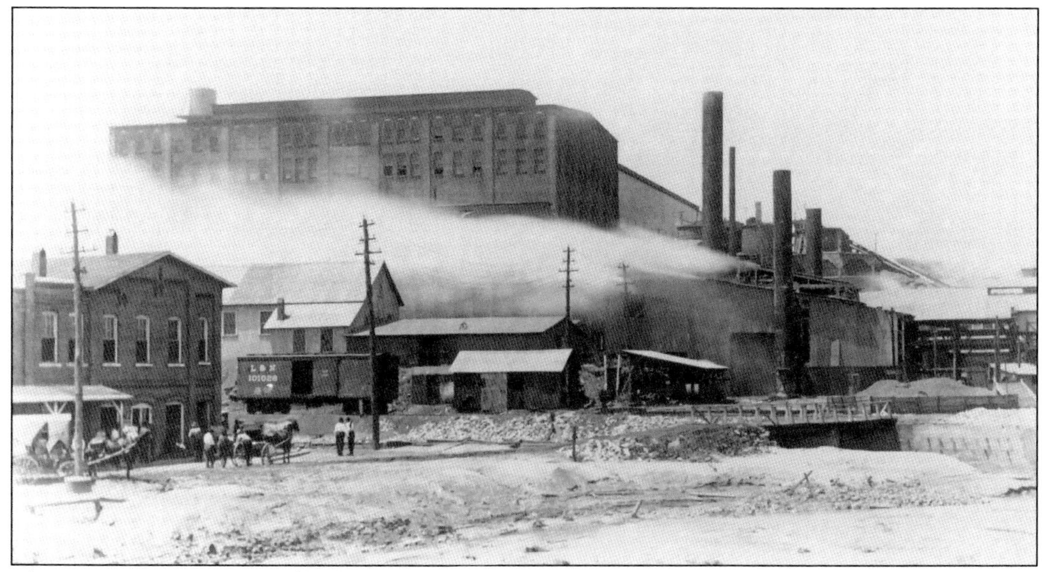

The DSC&I plant at Isabella dominates the skyline in this 1915 photograph, its facade partially obscured by escaping fumes from the smelter. Part of the company store can be seen in the left foreground. A train carrying express, freight, and passengers ran once a day between Isabella and the Copperhill depot. (Courtesy of John Roach.)

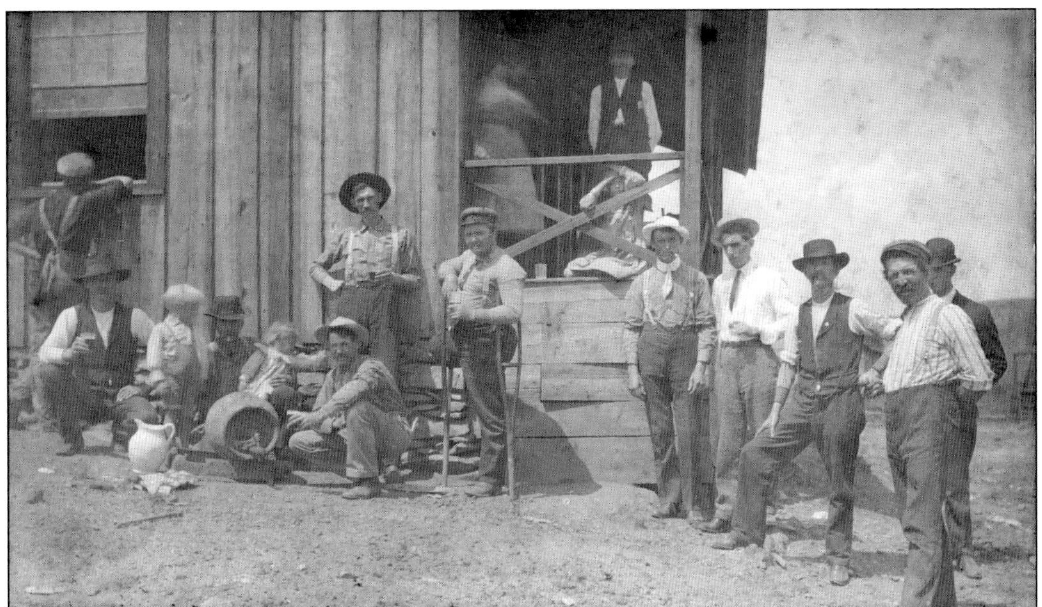

In the early days at DSC&I, happy hour soon became an established custom. The kegs of beer were brought into the Basin by express trains to Copperhill, then transferred to company trains for their journey to Isabella. The fourth man from right in this early 1900s photograph is DSC&I machinist Elijah "Lige" Roach. (Courtesy of John Roach.)

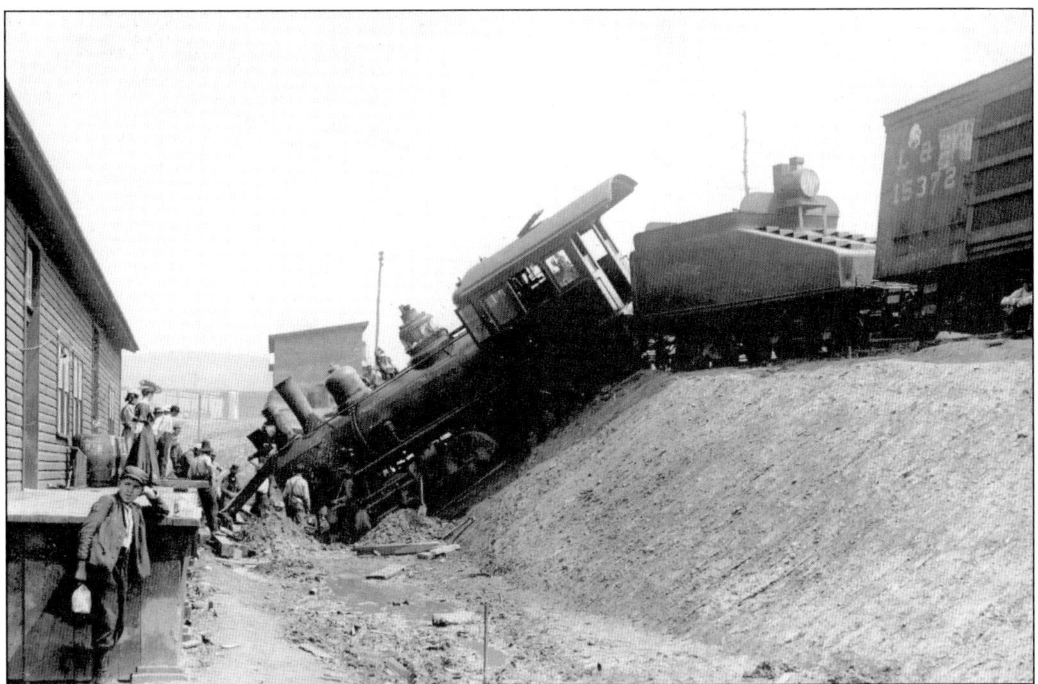

E.H. Westlake documented the aftermath of this Louisville & Nashville Railroad (L&N) locomotive accident in 1906. Although no location is specified in his notes, the topography, which includes a steep dirt bank similar to the one that can be seen in the photographs on the previous page, suggests that the photograph may have been taken at Isabella.

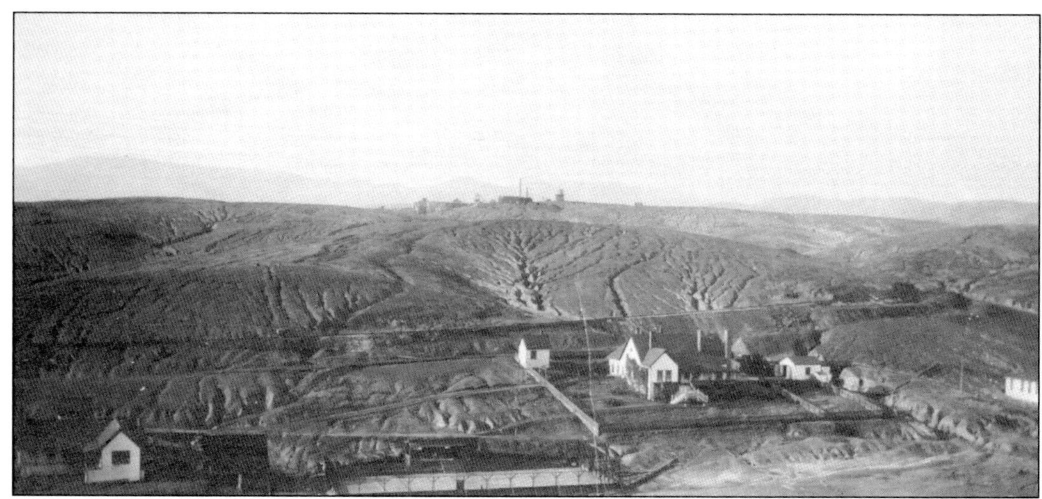

This photograph was taken looking north from the DSC&I staff house annex at Isabella toward East Tennessee Mine, on the horizon. It apparently dates from the tenure of general manager Charles W. Renwick, who left the company in 1918. The DSC&I tennis courts are in the foreground of the complex of company housing, and the only school is at far right. Older students went to school at Ducktown.

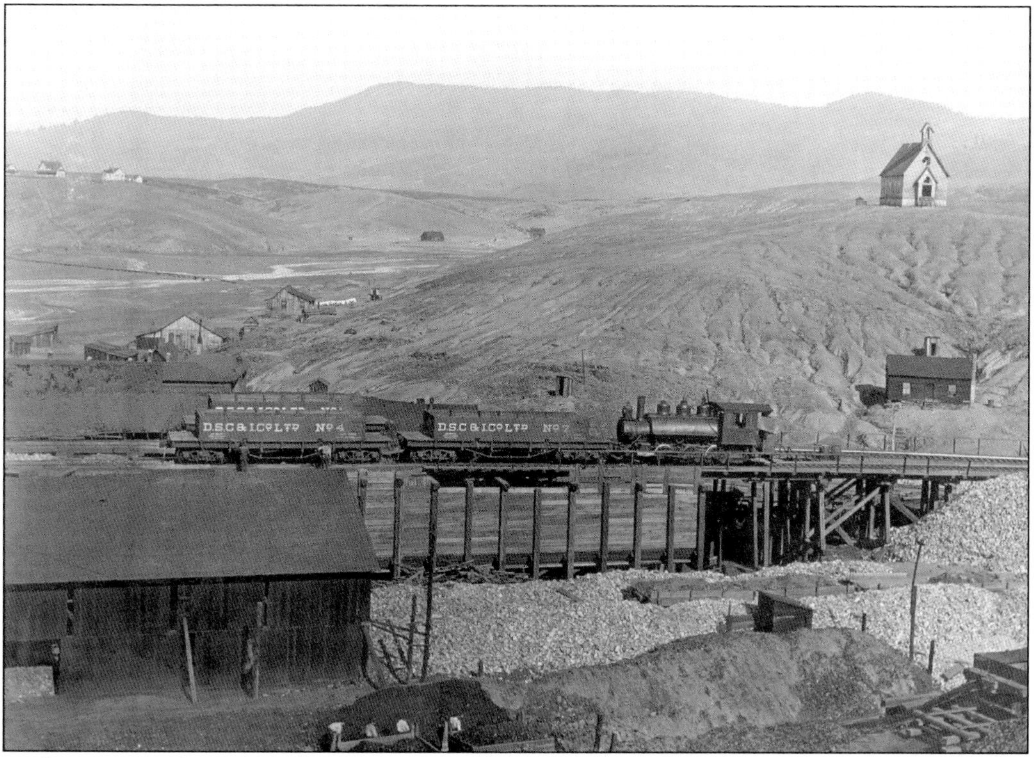

In this c. 1920 photograph of the DSC&I plant at Isabella, the old Union Church stands on a hill overlooking the railroad tracks. The church was built shortly after the Civil War by J.E. Raht of the Union Consolidated Mining Company, which had acquired Isabella Mine in 1866. For several years, it also served as a school. Above the front door is a spirit hole, intended to let the good spirits in and the bad spirits out.

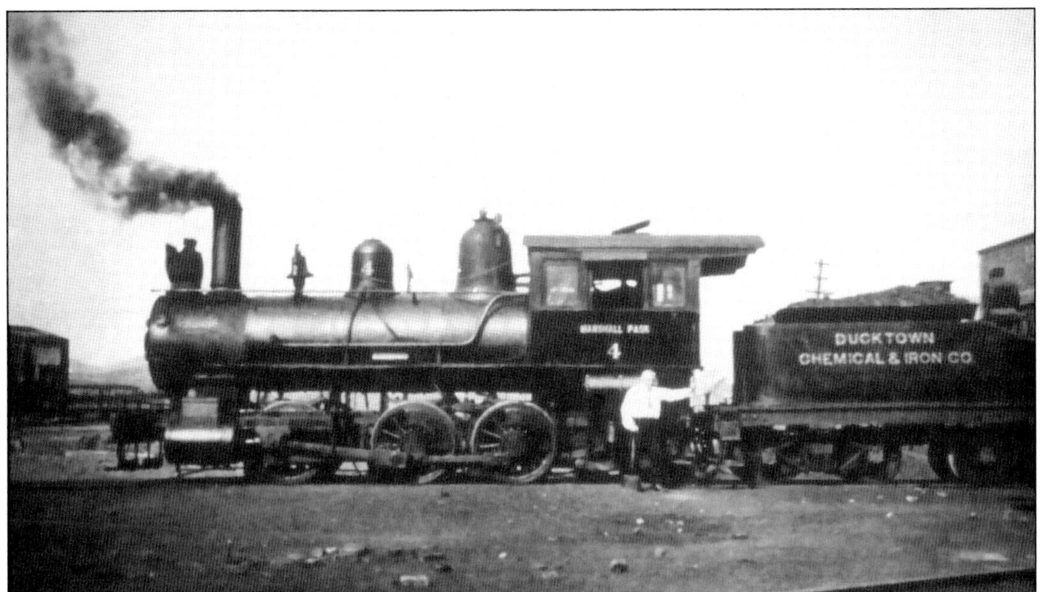

In 1925, following a period of financial difficulties, DSC&I was succeeded by the Ducktown Chemical & Iron Company (DC&I), which soon found itself facing financial difficulties of its own. In 1936, DC&I was acquired by Tennessee Copper Company's parent company, Tennessee Corporation, and became part of the TCC operations. This photograph was taken sometime during the 11-year tenure of DC&I between 1925 and 1936.

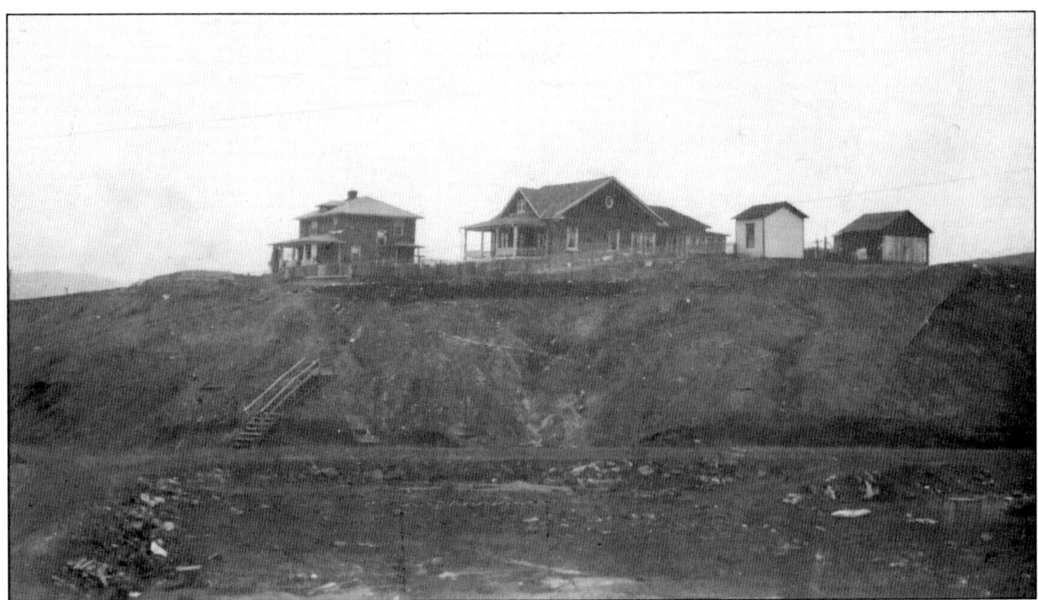

Many DSC&I senior management personnel retained their former positions when DC&I took over in 1925, although most were replaced after the transition to TCC management in 1936. In this photograph, the homes of assistant general manager P.B. Blakemore (left) and chief engineer H.E. Broughton stand on a hill near the Isabella plant. The photograph was apparently taken after the promotion of assistant general manager W.F. Lamoreaux to general manager in 1918.

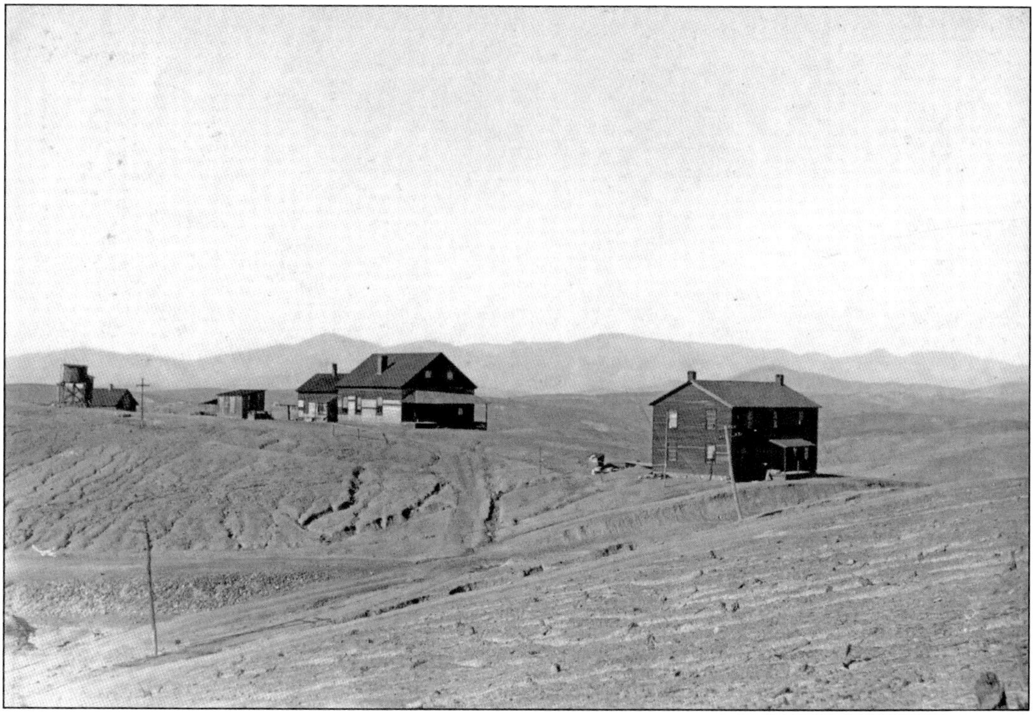

In 1899, the newly incorporated Tennessee Copper Company purchased the mining properties of the Pittsburgh & Tennessee Copper Company and moved temporarily into the former company's existing headquarters buildings on Polk County Hill. The general office and clubhouse in these two 1903 photographs by E.H. Westlake overlooked both TCC's Polk County Mine, acquired in the purchase, and DSC&I's adjacent Mary Mine, immediately to the northeast. The original Copperhill Post Office was located in the Mary Mine store. There were also several residences for managers and supervisors; these may have been the houses at the center of the photograph below.

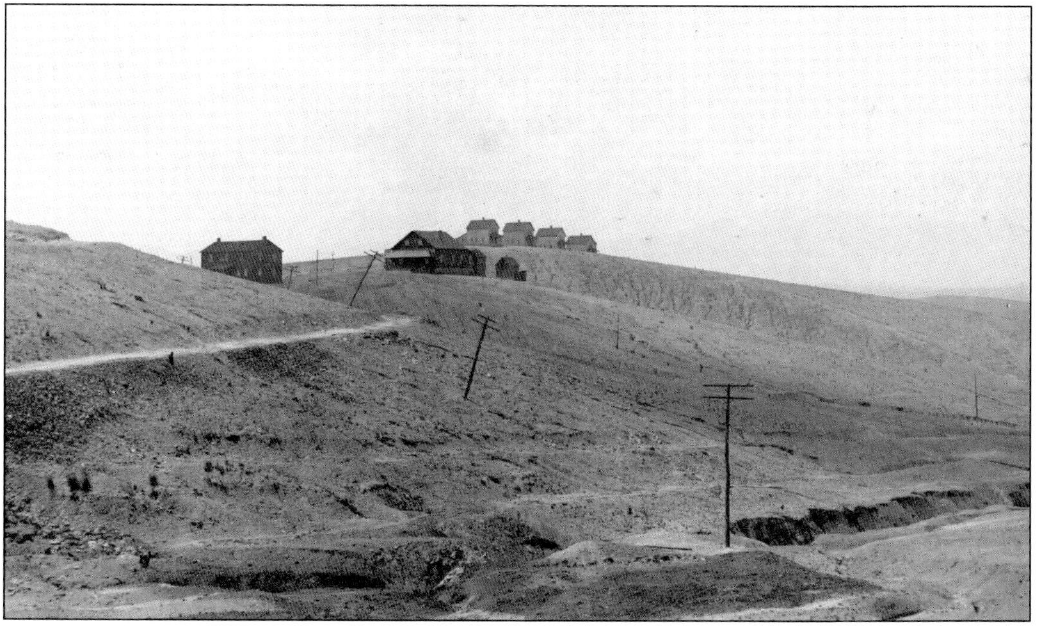

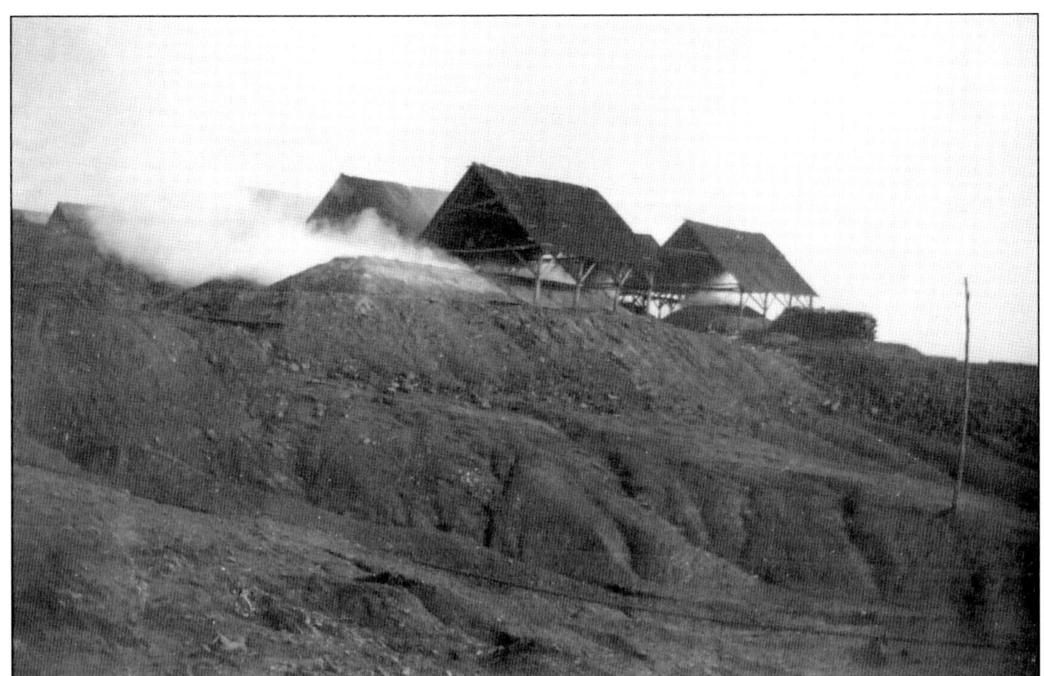

In the early days of mining, the first step in the smelting process was the roasting of the local copper ore in open sheds. The process required 65 cords of wood for each roast heap, took 60 to 90 days, and produced a dense sulfurous smoke that soon destroyed the surrounding vegetation. The four-acre roast yard at TCC's Polk County Mine was photographed in the early 1900s by Edward Oscar Boak (above) and E.H. Westlake (below.) The roast yard at Burra Burra Mine, a few miles away, extended across 14 acres. Open roasting was gradually discontinued after the advent of pyritic smelting in 1903.

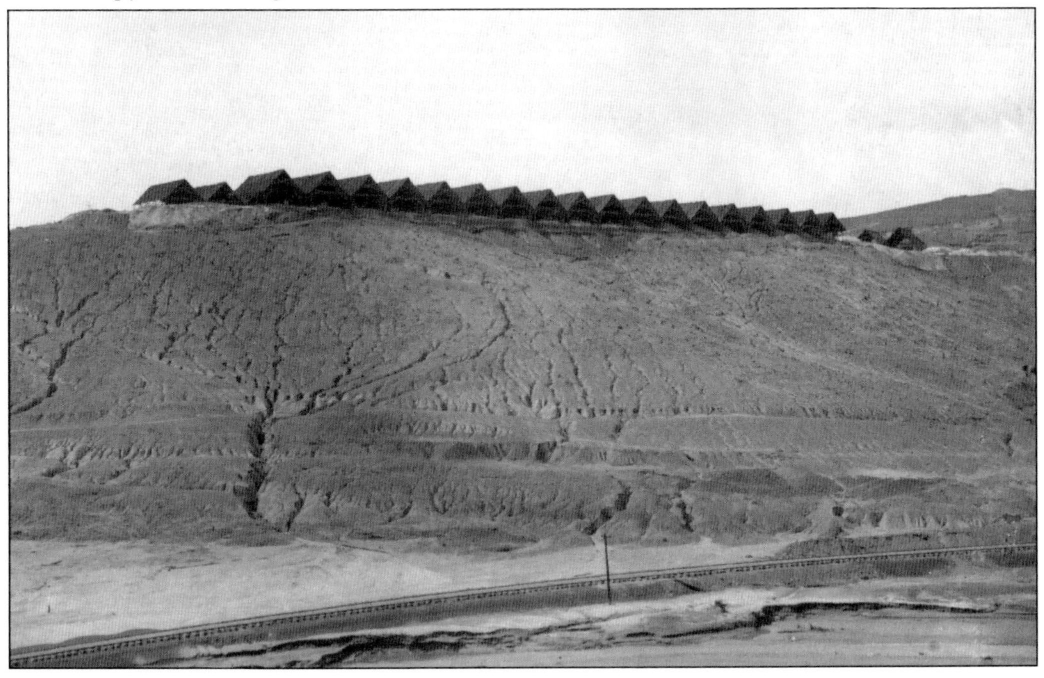

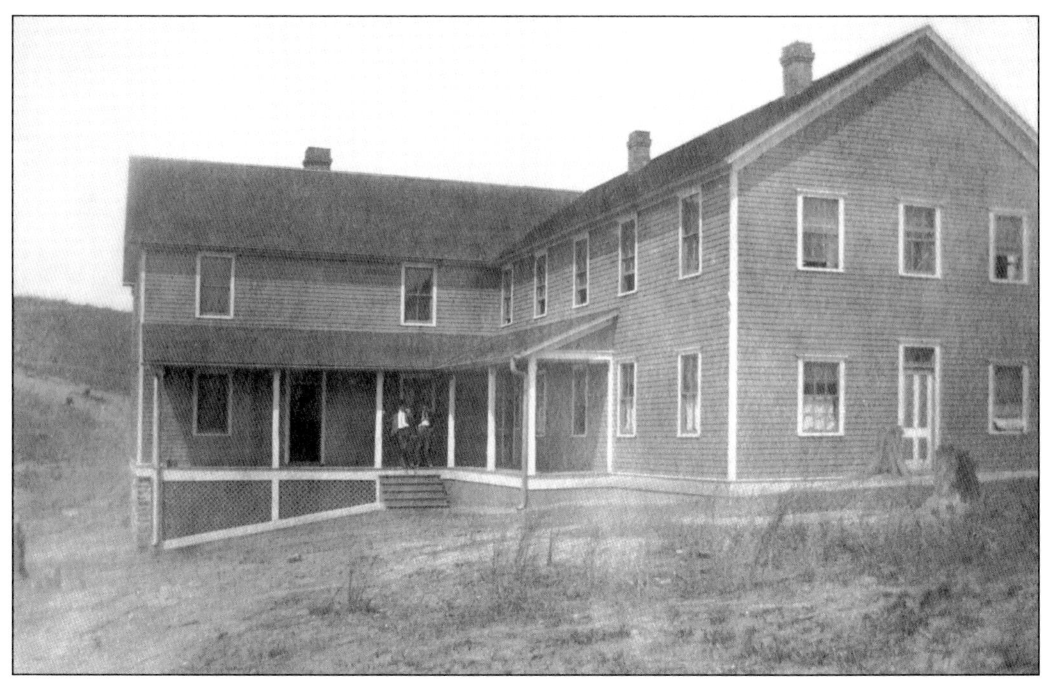

In 1900, the year after the incorporation of Tennessee Copper Company, an Iowa-born smelting engineer named Edward Oscar Boak was hired to help set up the company's first blast furnaces. Around 1902, he settled into the newly constructed Blue Goose residential hotel on Smelter Hill. By 1904, an addition to the original building had turned the Blue Goose into the L-shaped structure (above) photographed by Boak around that time. In 1907, TCC expanded the building again to help house the growing number of single employees; a later photograph (below) shows a second gabled wing that had converted the L-shaped building into a U shape. In the late 1940s, the original Blue Goose was replaced by a modern brick building at the same site.

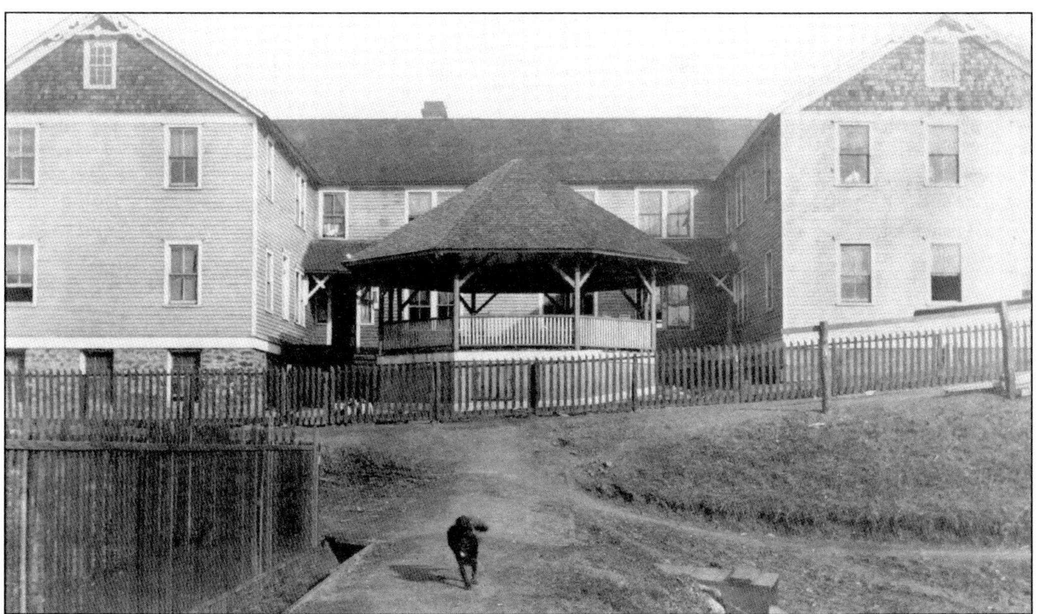

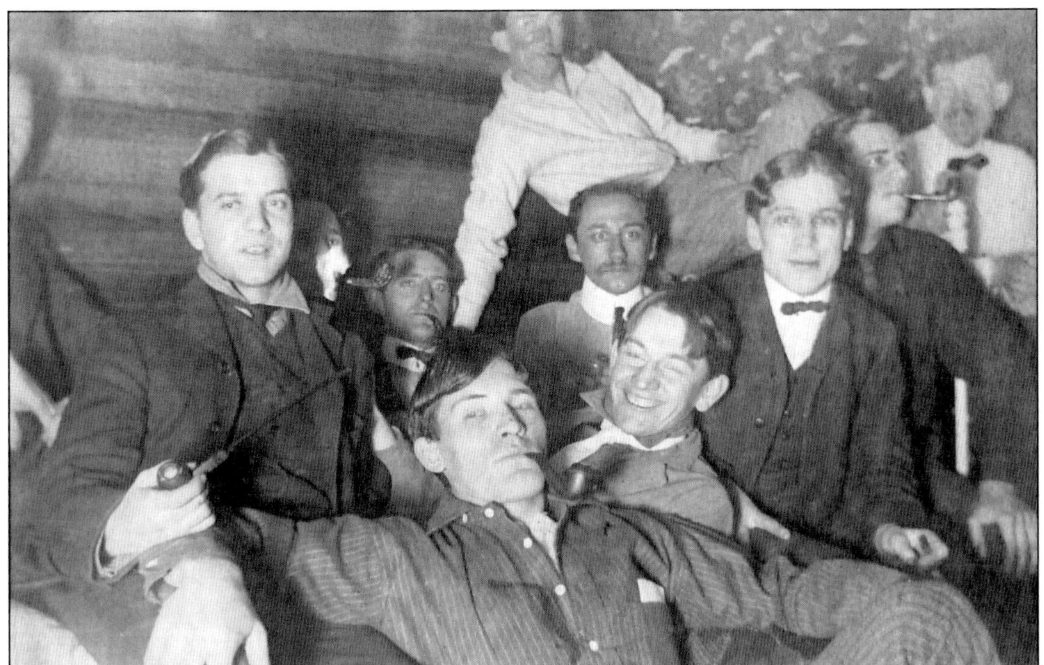

In Boak's photograph album, this image of his fellow boarders at the Blue Goose was labeled "The 'Blue Geese' on Barbour's bed." The majority of the hotel's residents at this point would have been young engineers and chemists from other parts of the country who had contracted for short-term employment at TCC before moving on elsewhere.

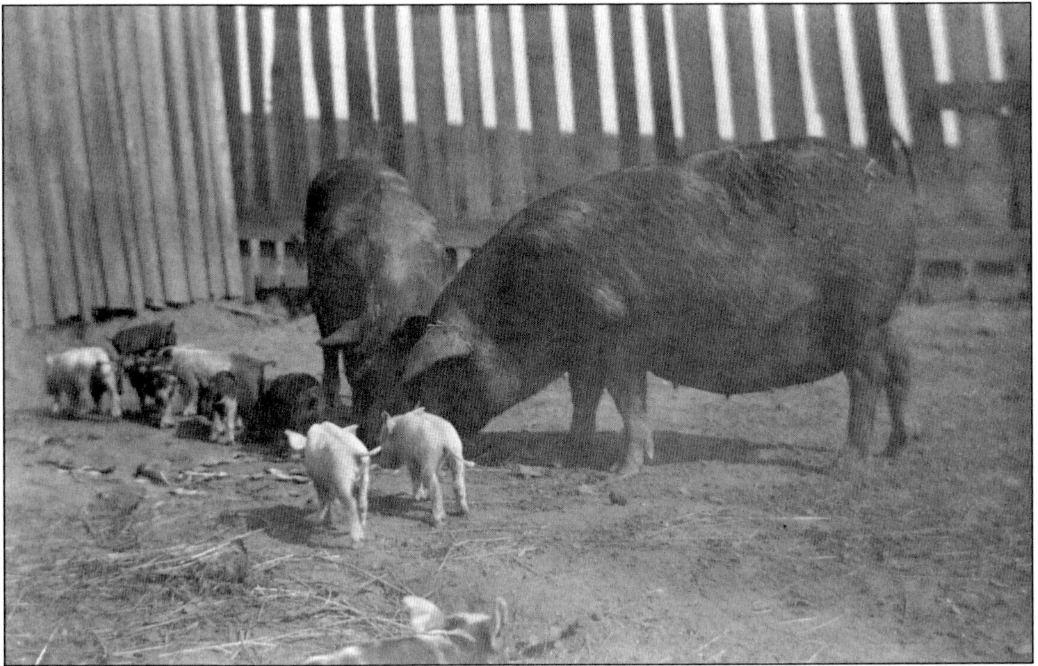

Boak's tongue-in-cheek label for this photograph was "The bill of fare at the Blue Goose." Since the Blue Goose served meals and the livestock was presumably destined for the table, his whimsical caption may have had a double meaning.

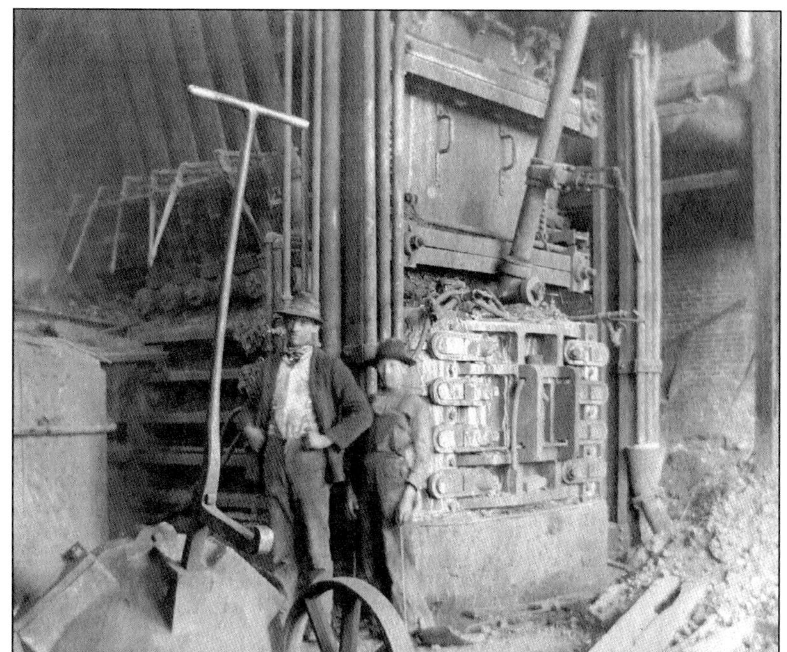

Edward Oscar Boak, who had come to the Basin in 1900 to help TCC set up its smelting operation, took this early photograph of two workers at one of the company's first blast furnaces. By 1906, TCC had seven of these furnaces in operation, four of which were the largest copper blast furnaces in the country.

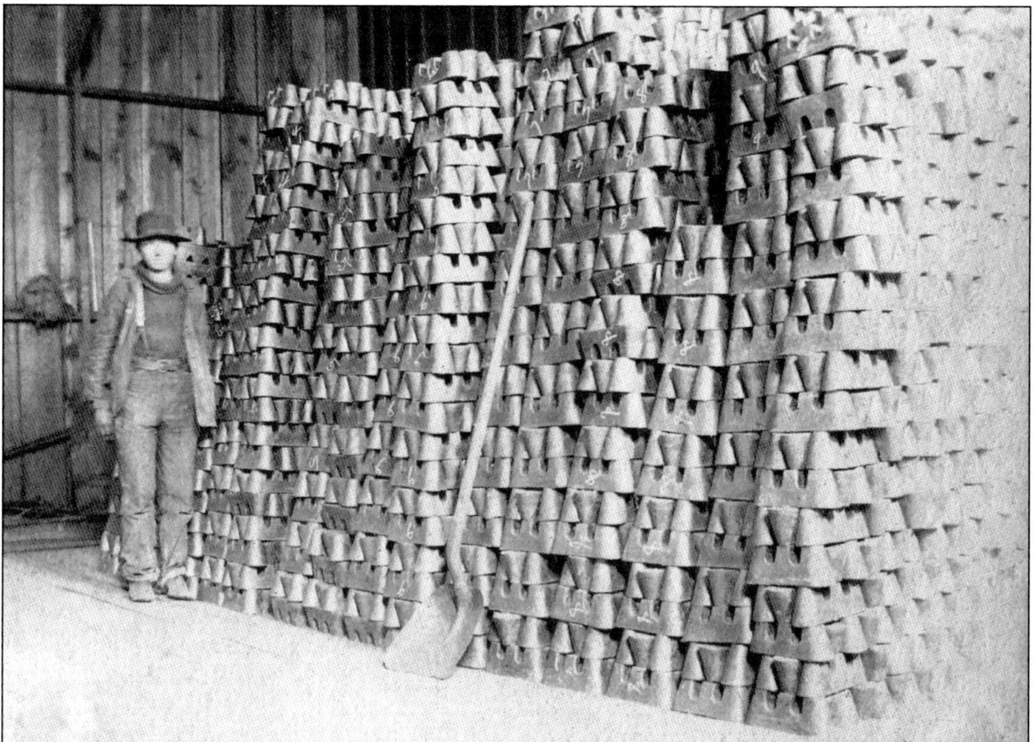

This unidentified worker is standing next to 300,000 pounds of copper ingots, which had been formed by pouring molten copper into molds. Edward Oscar Boak apparently took this photograph between 1902, when the company installed a reverbatory refining furnace, and 1904, when on-site refining at the TCC smelter was temporarily discontinued.

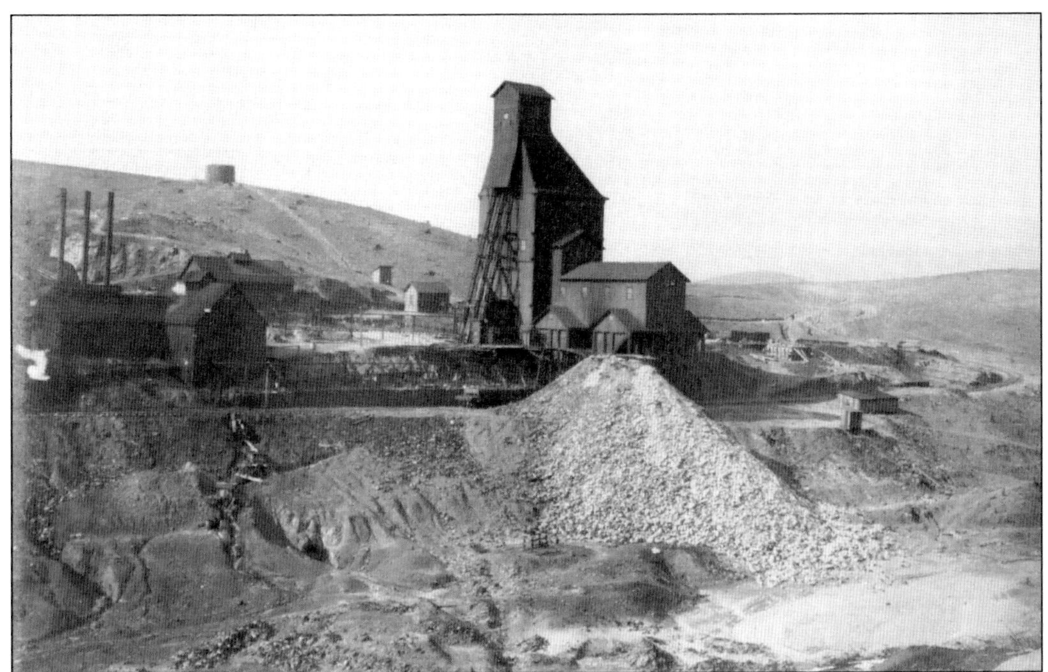

London Mine, first opened in 1853, was photographed by Edward Oscar Boak shortly after its reopening by Tennessee Copper Company in 1901. The mine was permanently closed in 1926 and became the site of the TCC flotation mill, which separated the minerals found in the raw ore. The large pile in the foreground is waste rock from the mining operation.

This photograph of the slag pot and locomotive at TCC's Copperhill plant was taken by Edward Oscar Boak between 1900 and 1905. The black slag, a waste product from the blast furnaces, was tapped into pots like these for disposal. The high black bluff that still stands adjacent to the Copperhill plant site was created over the years by the dumping of the molten slag.

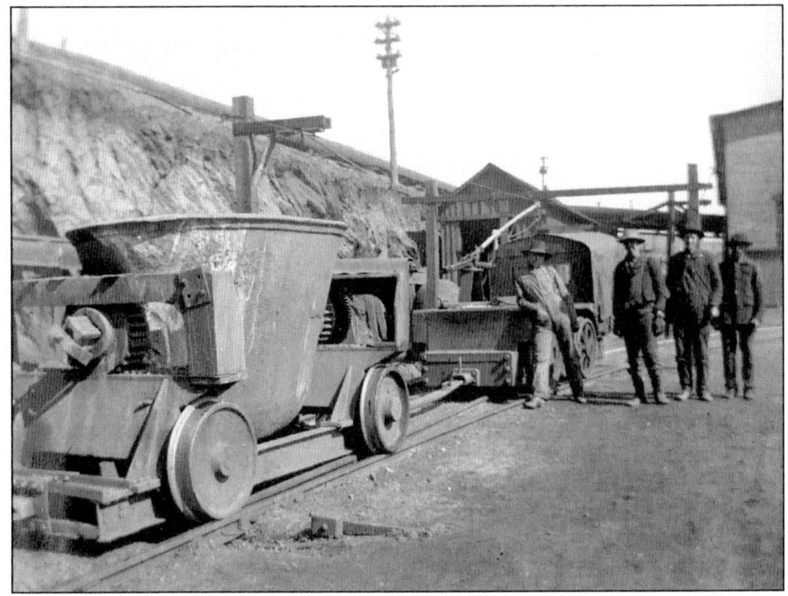

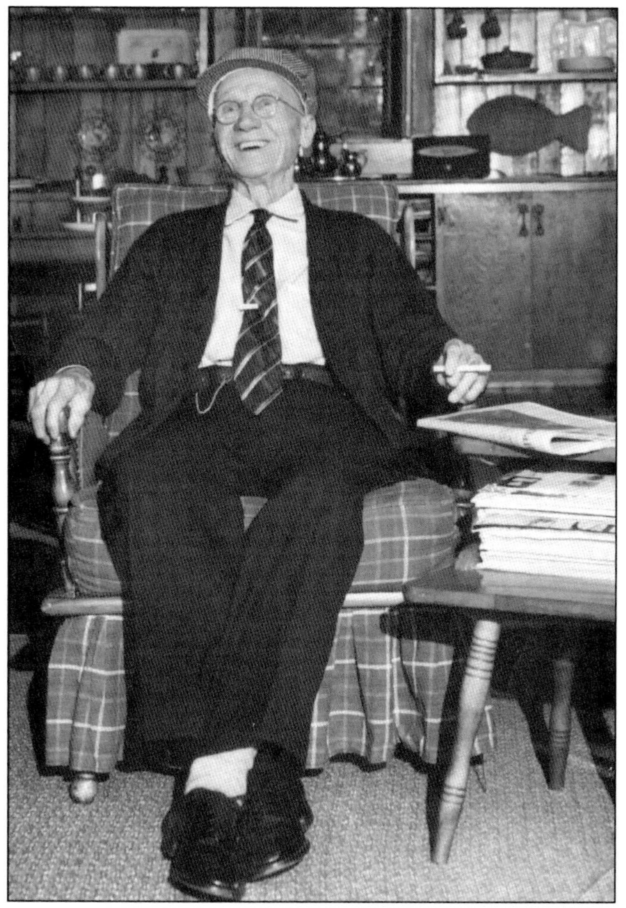

When the first trainload of copper was shipped from Tennessee Copper Company on November 7, 1901, engineer Chesley Jones was in the driver's seat. Inevitably, his coworkers christened him "Casey," a nickname he carried with him when he left TCC a few years later. After working for more than two decades as a locomotive engineer in the North and Midwest, he changed careers and became a millwright. The photograph at left was taken in 1958 at the home of his daughter and son-in-law Imah and Fred Shelton on Blue Ridge Lake. The train that hauled that first load of copper would have been pulled by a steam locomotive much like the Baldwin 0-6-0 below, photographed in 1905 by E.H. Westlake.

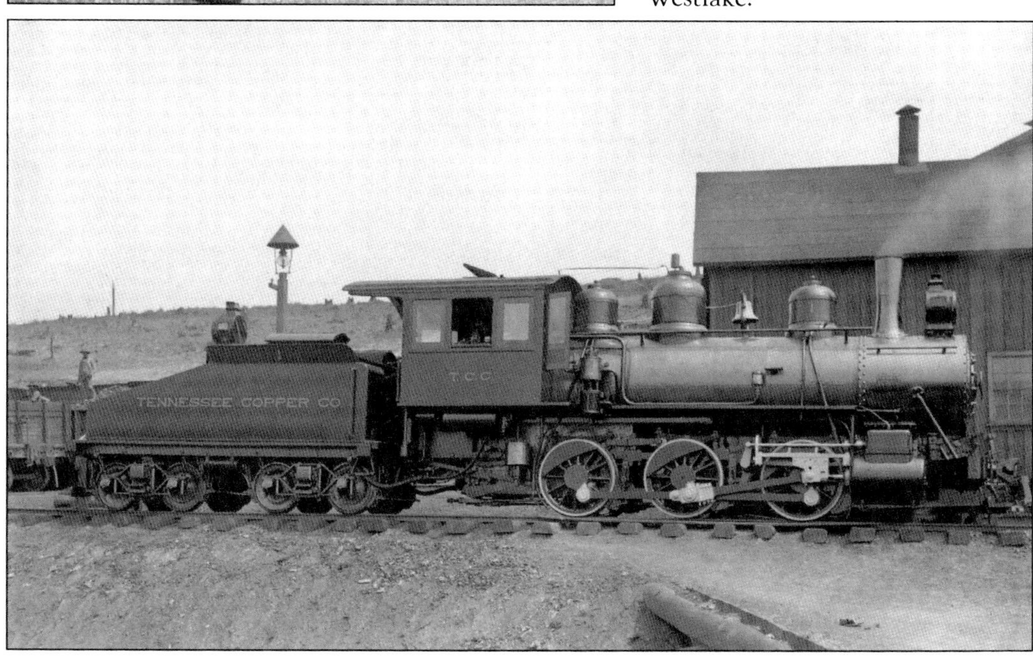

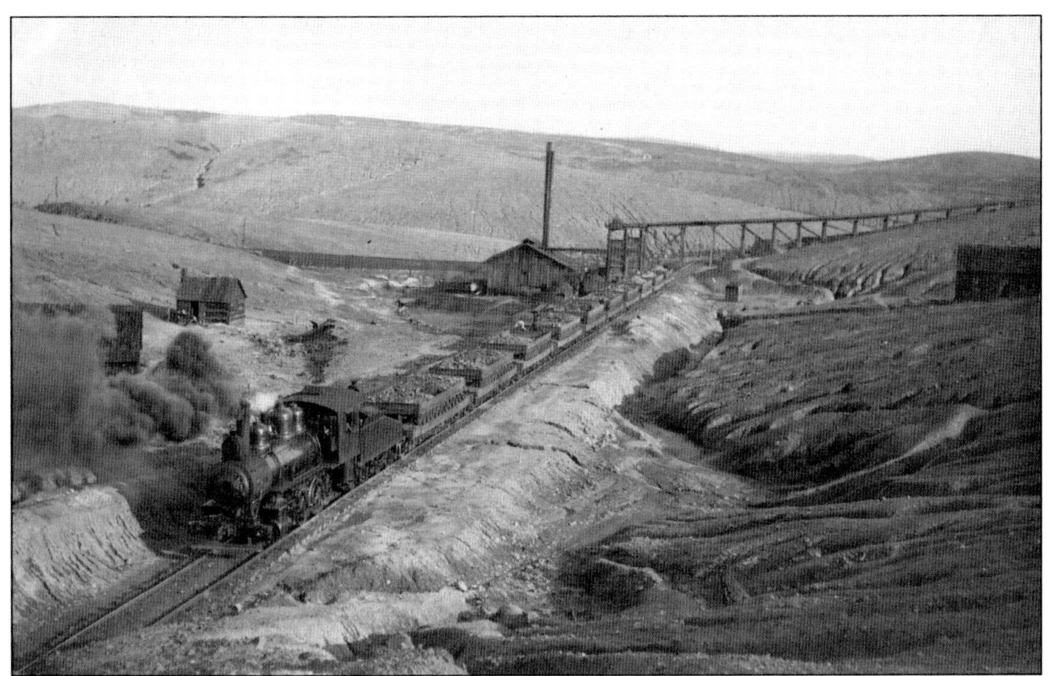

Above, a TCC ore train is loaded and headed toward the TCC plant at Copperhill. Somewhat curiously, this photograph appears to have been taken at DSC&I's Mary Mine, and the TCC train appears to be on the DSC&I rail line, which was separate from the TCC line. The tracks of the two companies ran parallel at some points but diverged again just north of the TCC smelter (below). The TCC line curved around to the east side of the smelter, and the DSC&I line joined the L&N Railroad tracks farther west at Isabella Junction, which is marked by the white building at center. Both photographs were taken in 1905 by E.H. Westlake.

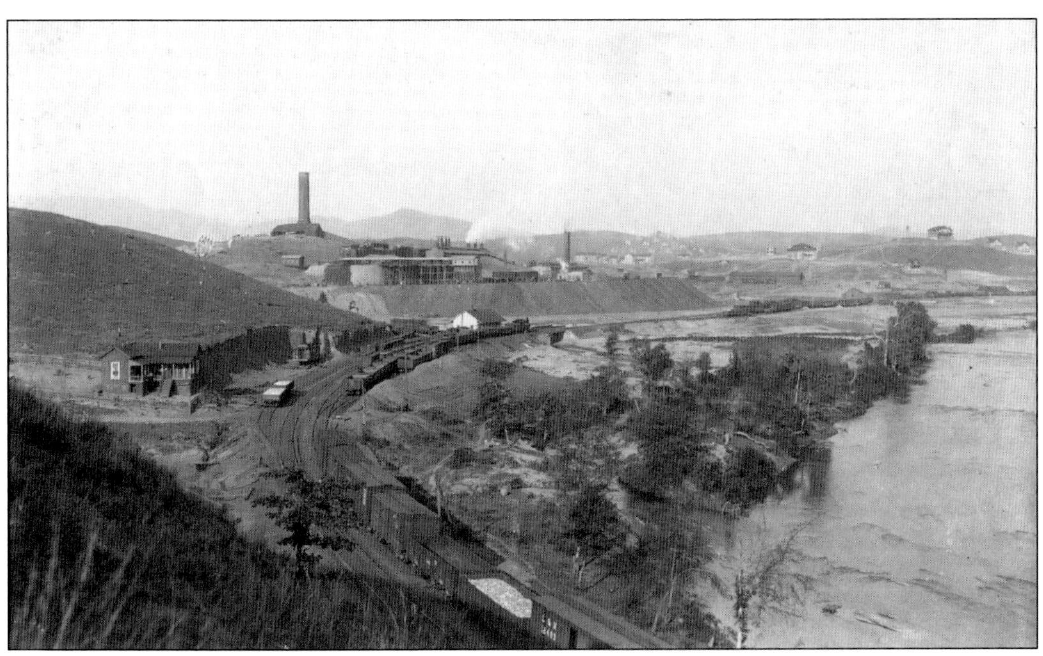

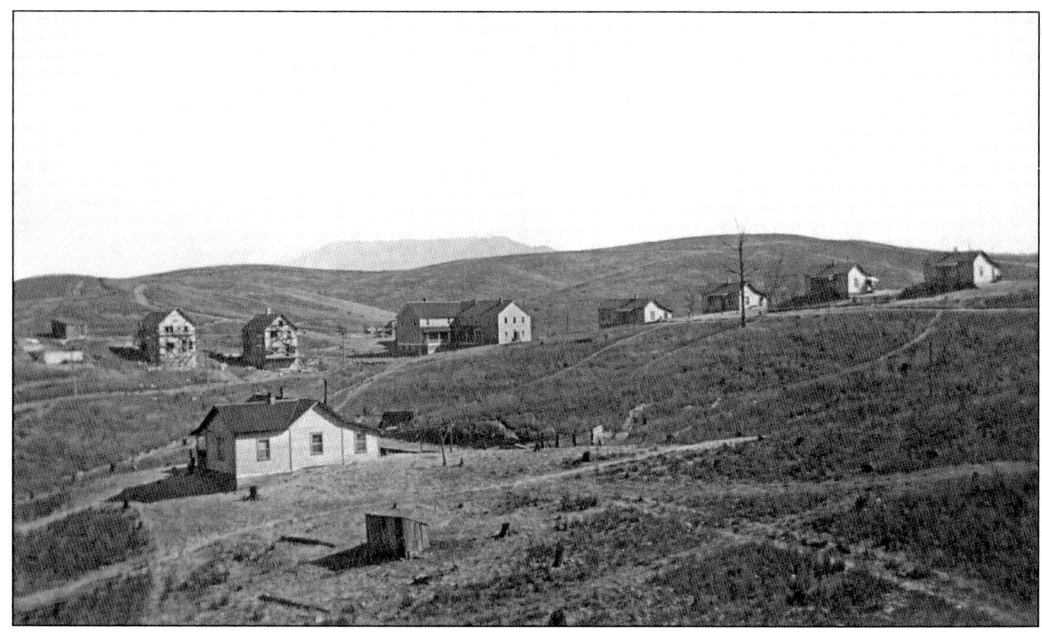

These two photographs, probably taken no more than a few months apart, document the pace of construction on Smelter Hill after Tennessee Copper Company began building homes for its managerial staff in 1903 and 1904. At the center of the photograph above, taken by Edward Oscar Boak, is the Blue Goose after the addition of its L-shaped wing. Scaffolding can still be seen on the two houses at left, and a row of more modest houses has been built on the right. By the time E.H. Westlake photographed the same scene in 1904 (below), the houses at left had been completed, additional houses were under construction, and the Blue Goose had apparently been painted white. By the end of 1904, the records show that TCC owned about 60 houses intended for its employees.

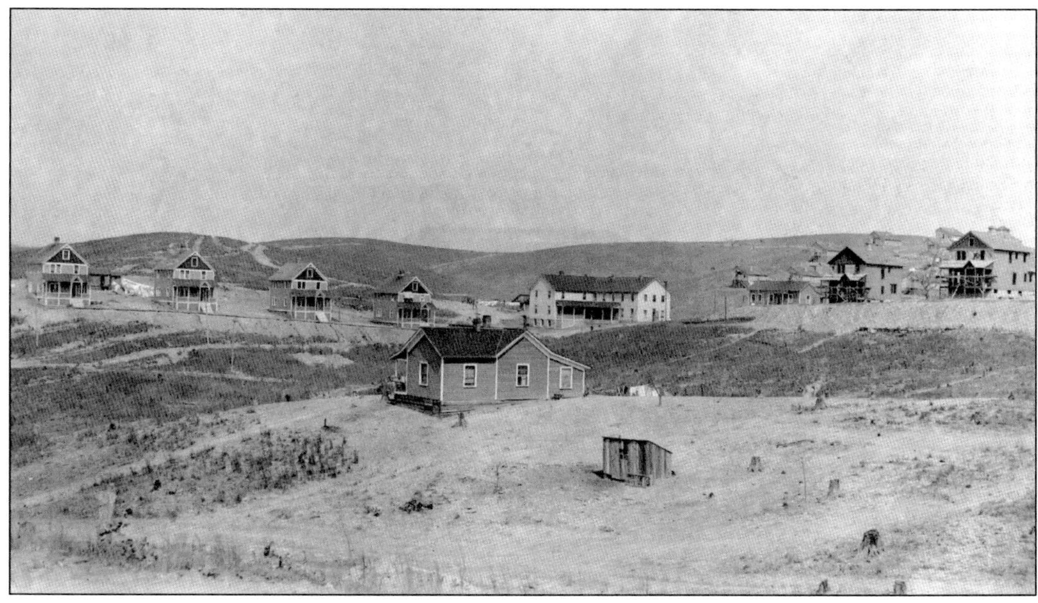

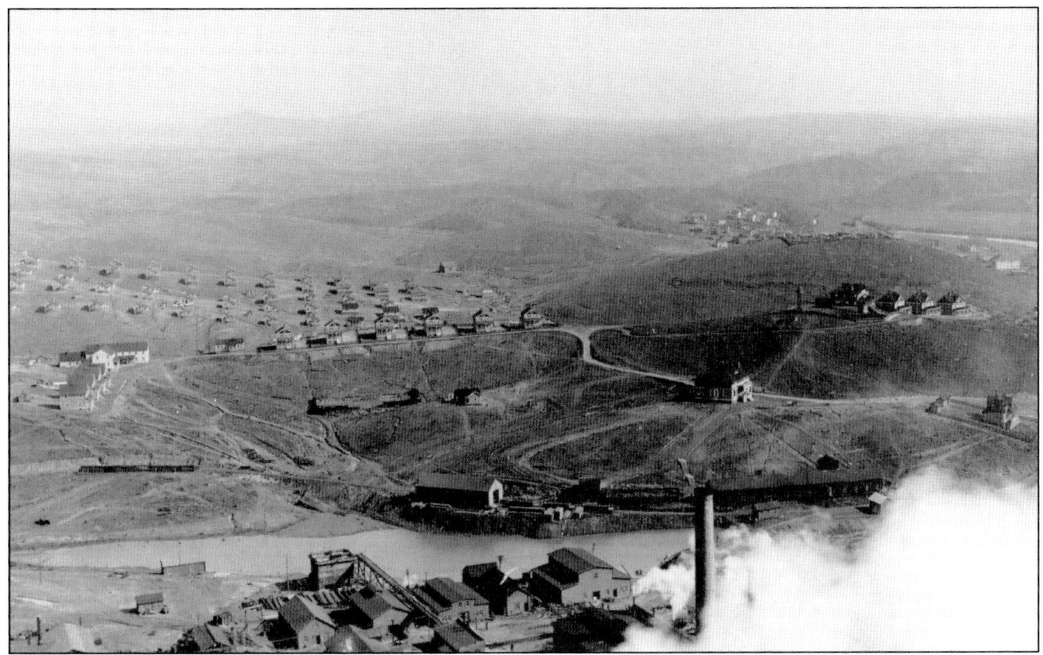

In this 1906 E.H. Westlake photograph looking east from the TCC smelter, the Blue Goose is on the far left, beyond the row of four identical company houses. Just past the ridge are the smaller workers' houses of Newtown. The general office is the building at center right, and the homes of TCC management officials are on the hill at far right. Part of the town of McCays can be seen in the distance at right.

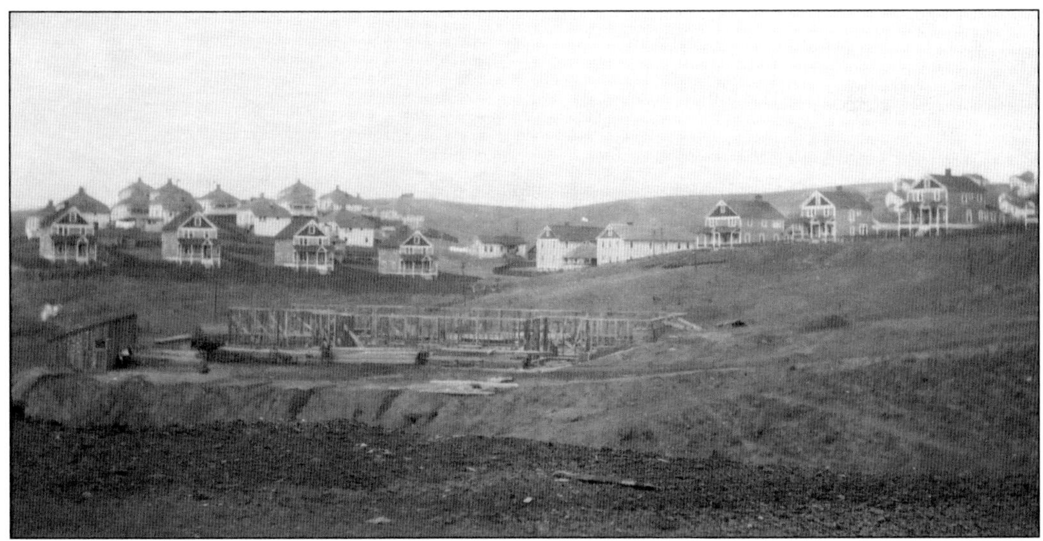

In this 1907 photograph, Smelter Hill has continued to expand to meet the demand for housing of TCC's upper-level staff. A second wing and a gazebo-style porch have been added to the Blue Goose, just right of center. In the foreground, construction has begun on the Cowanee Club, which had been organized a few years earlier but had not previously had its own facilities.

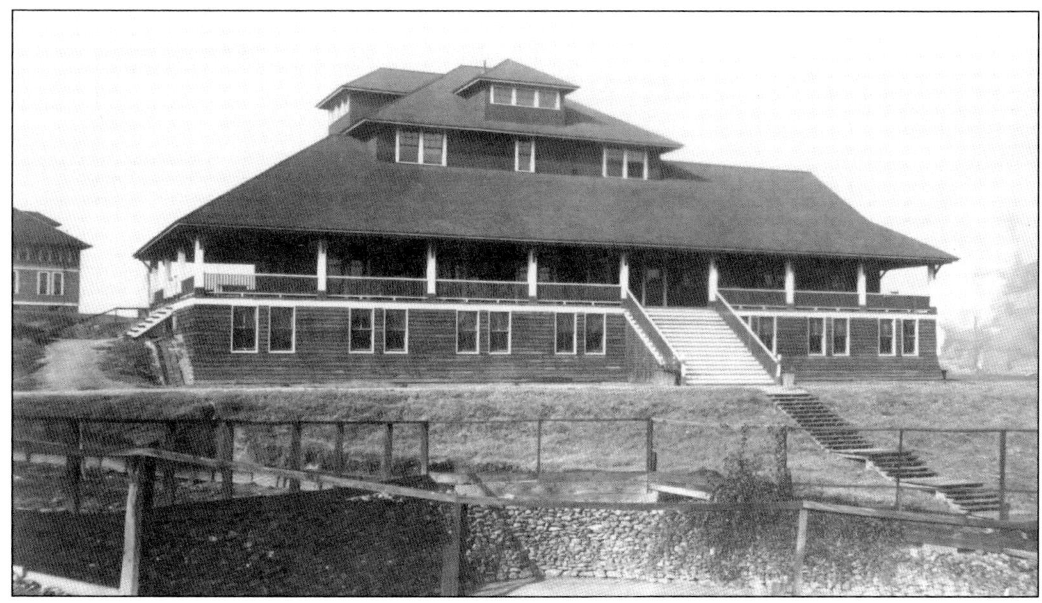

In the early 1900s, members of the TCC engineering staff organized the Cowanee Club for the purpose of arranging social functions for themselves and their guests. The club initially met at the Blue Goose and also maintained quarters in the general office building. In 1907, with help from TCC, the private club began construction of its own building (above) with tennis courts, a swimming pool, a gymnasium, a billiard room, a bowling alley, a kitchen, and a large main hall that could be used for dances and receptions. The building also had four furnished upstairs bedrooms for use by visiting company officers and directors. The 1916 photograph below, from the files of E.H. Westlake, shows the club's location relative to TCC's employee housing on Smelter Hill.

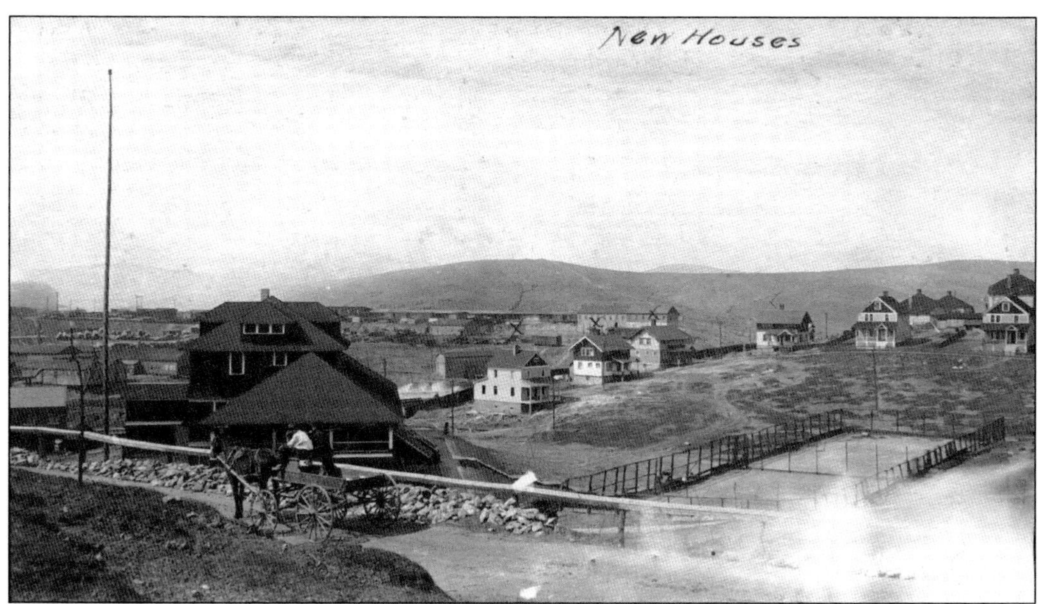

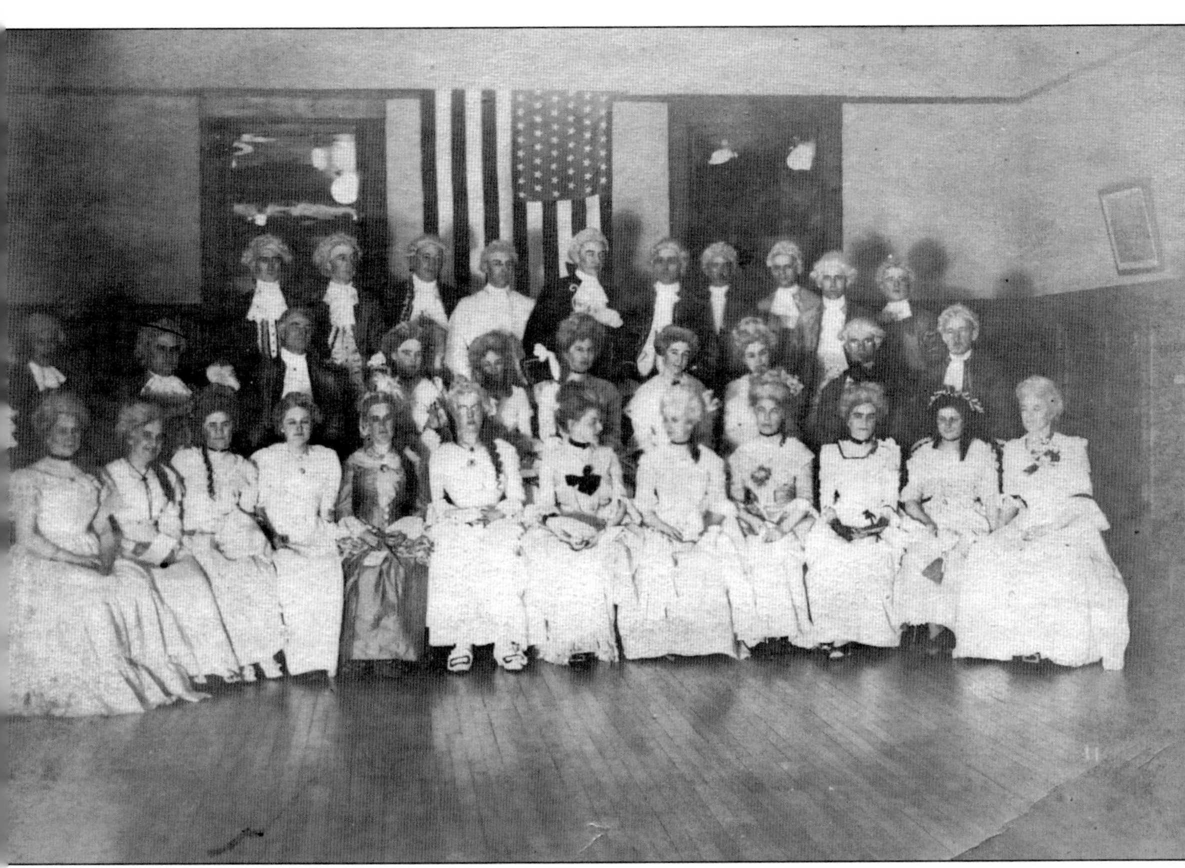

On February 22, 1910, in honor of Washington's Birthday, the Cowanee Club held a Colonial dance at its recently completed clubhouse. In attendance were, from left to right, (first row) Marcelina Daniel, wife of Preston Daniel; Mary Slaughter, wife of Benjamin Slaughter; Carrie Janney, wife of Walter Janney; Jennie Guess, wife of George Guess; ? Scott; Elizabeth Caine, wife of Milton A. Caine; ? Dryden, wife of D.F. Dryden; ? Graves; Marion Hubbard, wife of Preston Hubbard; ? Lepley; and ? Pomroy; (second row) stenographer Preston Daniel; draftsman Walter Janney; engineer Benjamin Slaughter; ? Fogh; ? Cavers, wife of T.W. Cavers; Mae Walker, wife of Dr. Evahn Walker; Marion Emmons, wife of N.H. Emmons; TCC general manager N.H. Emmons; Edna Brodie, wife of John Brodie; ? Fairline, wife of A.M. Fairlie; A.M. Fairlie; and ? Johnson; (third row) chemist T.W. Cavers; smelter superintendent George Guess; purchasing agent John Brodie; mine superintendent Milton A. Caine; draftsman Preston Hubbard; acid plant superintendent D.F. Dryden; draftsman Guy Baltzell; ? Hardie; Dr. Evahn Walker; and chemist W.R. Yonge. In 1916, Milton A. Caine was named assistant general manager at TCC, T.W. Cavers was named smelter superintendent, and A.M. Fairlie was named superintendent of the acid plant.

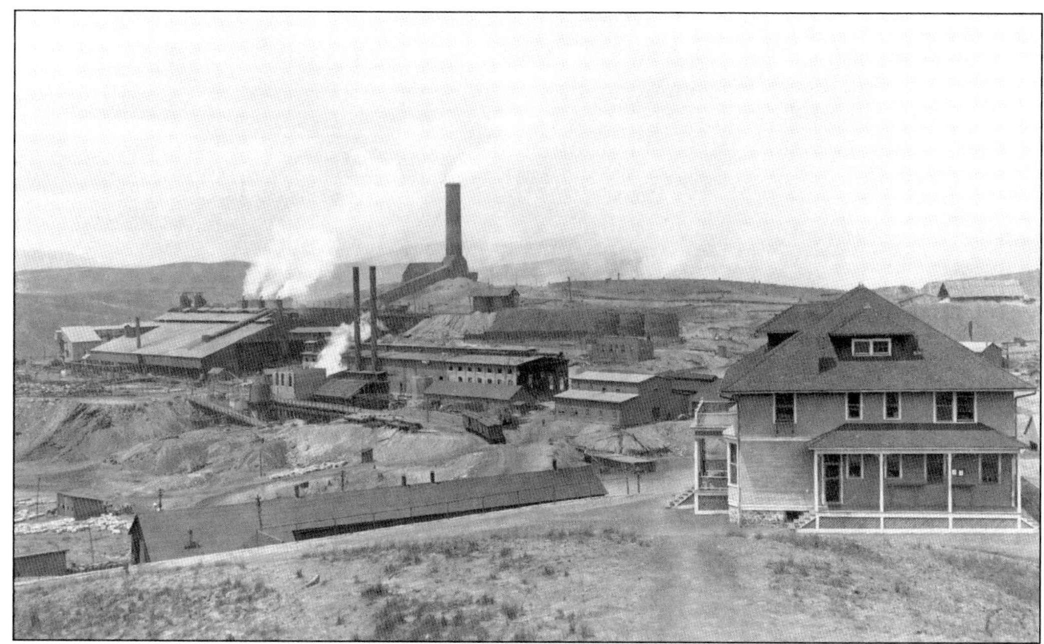

In 1904, TCC completed the construction of its new general office building, which overlooked the smelter at McCays. By the time E.H. Westlake took these two photographs in 1905 (above) and 1906 (below), the company was already dealing with legal issues caused by smoke damage from the smelter, and plans were being made to build a new 325-foot smokestack to the west of the existing 150-foot stack. It had also become clear that the building of a sulfuric acid plant would be necessary in order to contain the noxious sulfur fumes, deal with the pending smoke injunctions, and continue smelting operations at McCays. Plans for the acid plant were on the drawing board by mid-1905.

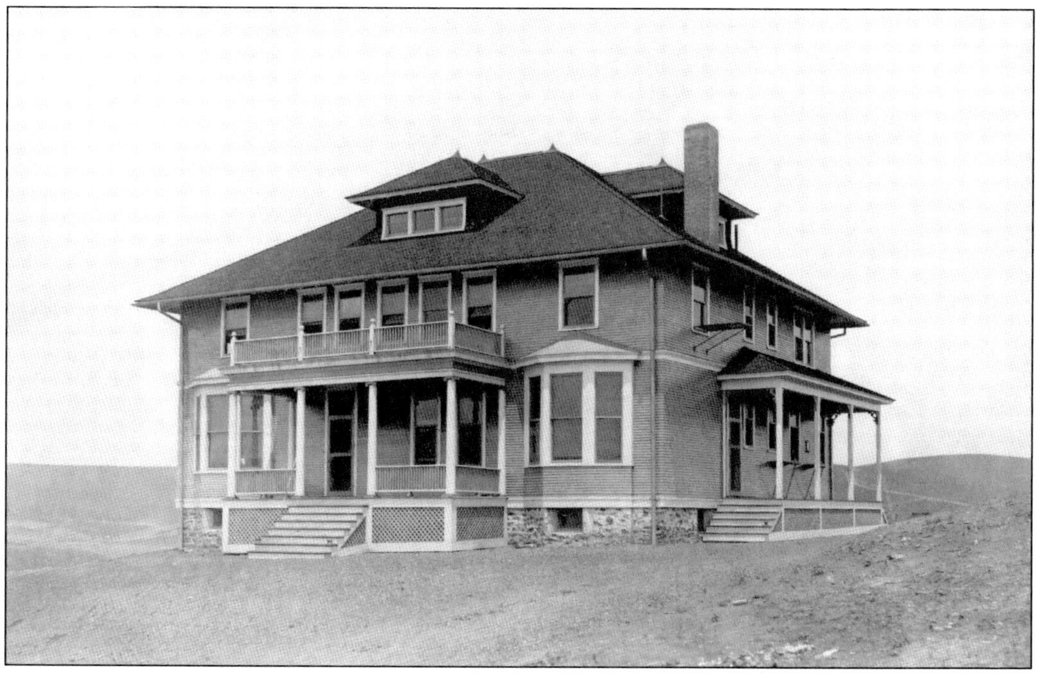

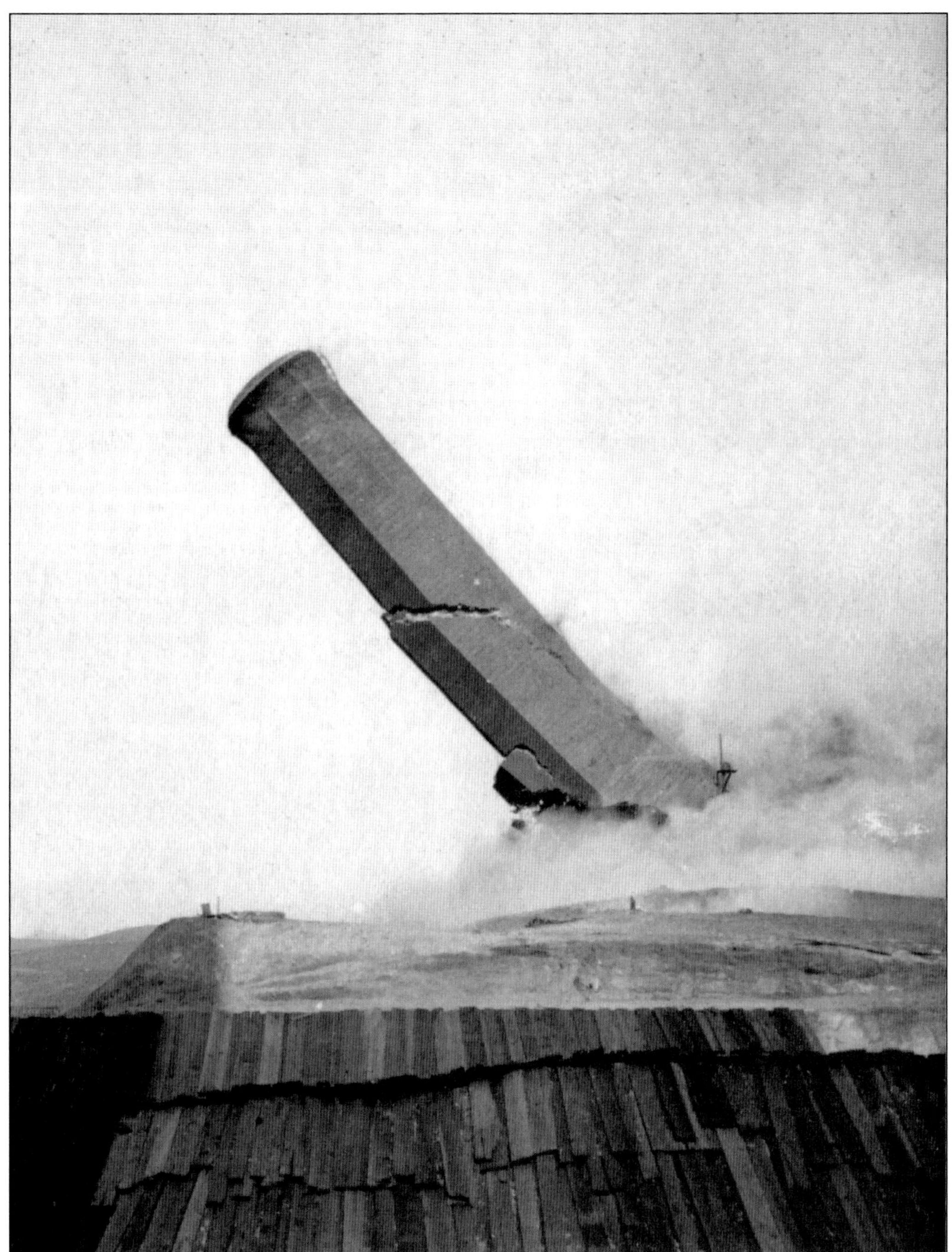

In early 1906, TCC began construction of the No. 1 acid plant, which would be the first ever to manufacture sulfuric acid from smelter furnace gases. It was a bold experiment, but it was one that the company considered vital to its survival. To make room for the plant, it was necessary to level the entire crown of the hill where the original 150-foot smelter stack stood. E.H. Westlake and his camera were there to record the drama of the actual moment when the smelter stack came down in 1906.

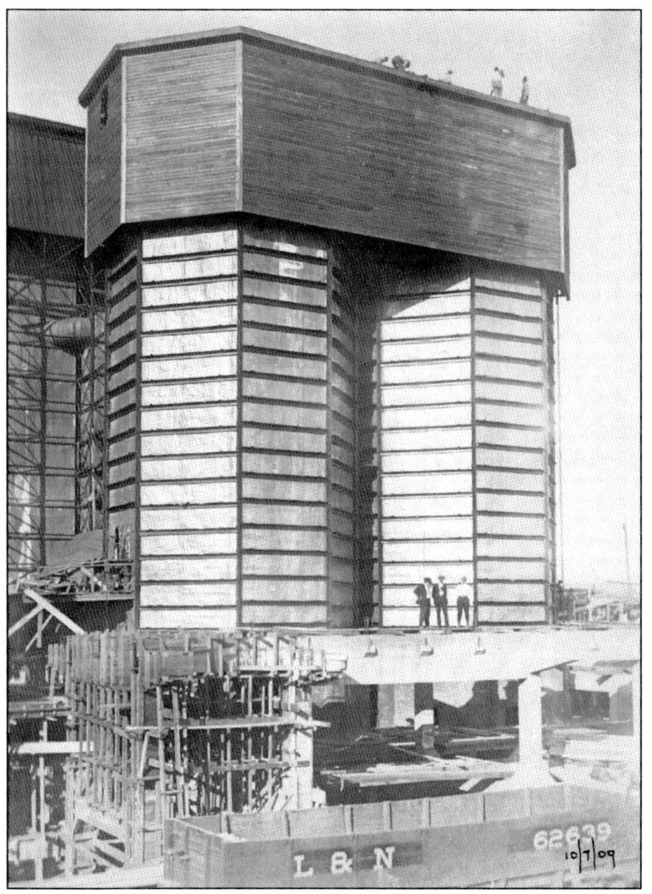

To the surprise of those who had originally considered it merely a means of dealing with the smoke damage suits caused by the escaping sulfur fumes, the sulfuric acid plant became an unexpected source of profit for TCC. In 1909, construction began on an addition that would double the size of the existing plant. The photograph at left was taken in October 1909, two months into the project. In the photograph below, taken in August 1910, the addition is nearing completion. When the expanded plant began full operations in December 1910, it was the largest sulfuric acid plant in the world.

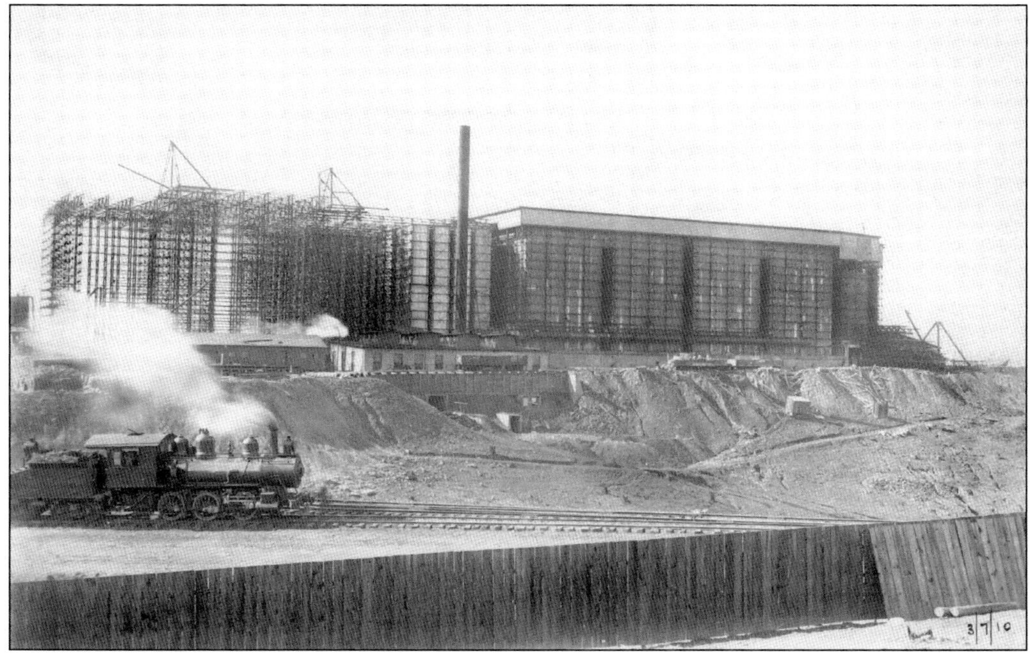

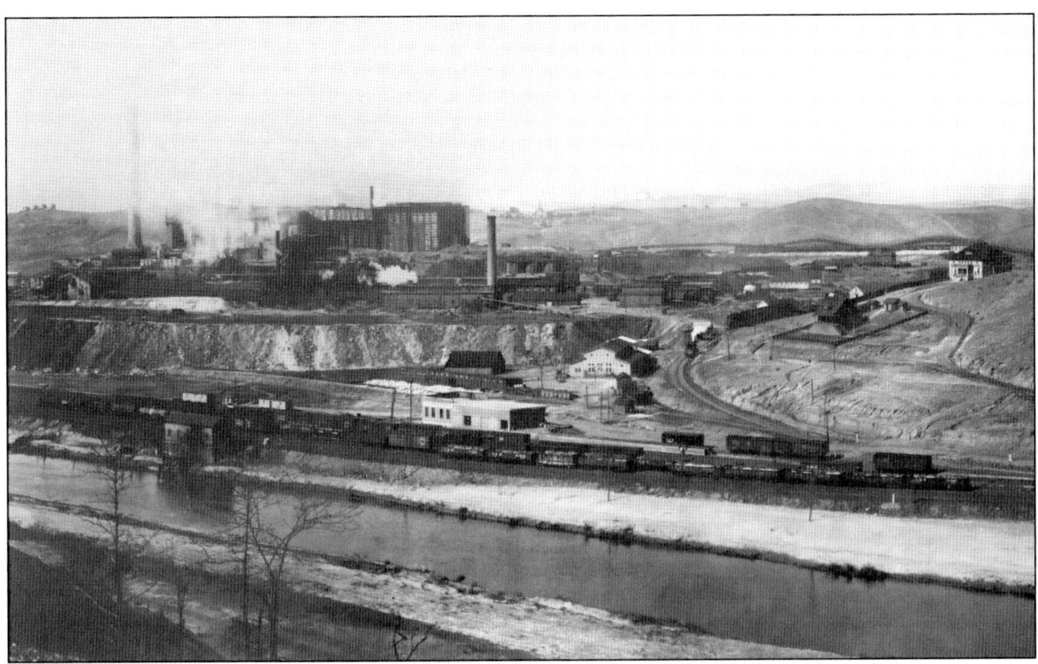

By 1911, when the photograph above was taken, all available ore from the Burra Burra and London mines was needed to keep sulfuric acid production at TCC up to full capacity. Production was temporarily interrupted that year by a fire that closed down operations for three months. By September, however, the plant was back in operation and producing 500 tons of acid a day. By spring of the following year, its daily production had grown to 700 tons. The TCC general office can be seen at right. Below, an early postcard image provides a closer view of the completed expansion.

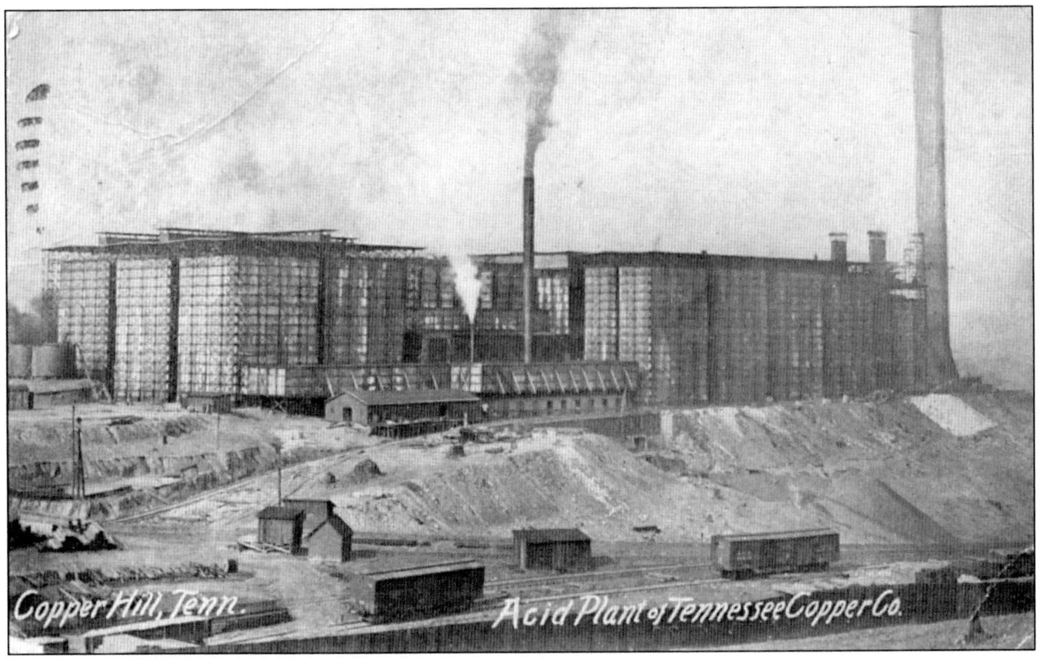

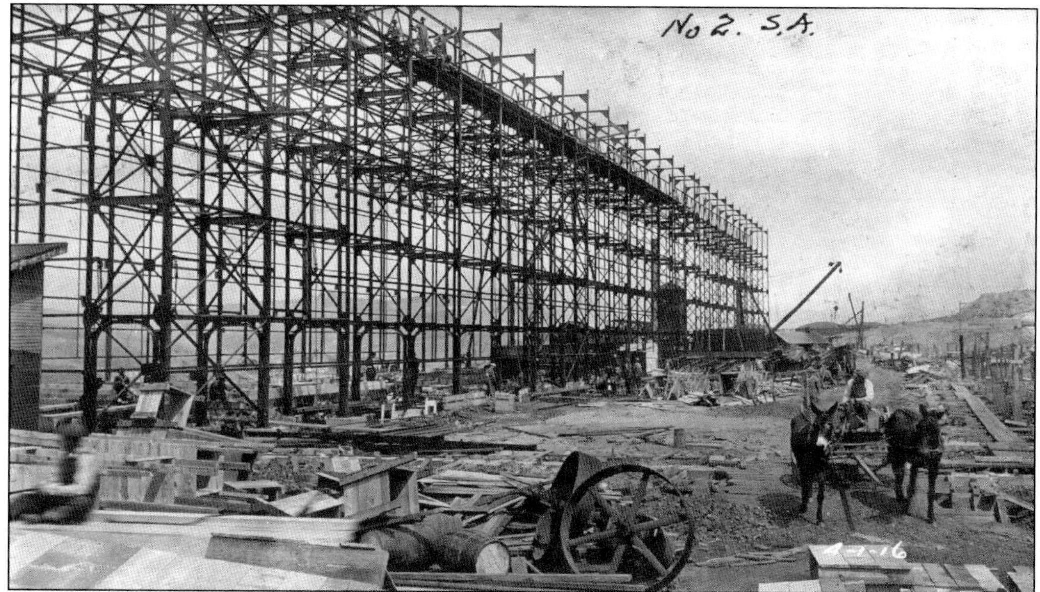

In 1916, due to the increasing demands for sulfuric acid during World War I, TCC began construction of its No. 2 acid plant. This photograph was taken in early April of that year, and the plant was put into operation in late July. At the same time, a considerable amount of work was being done on other TCC facilities, creating jobs for both construction workers and permanent employees.

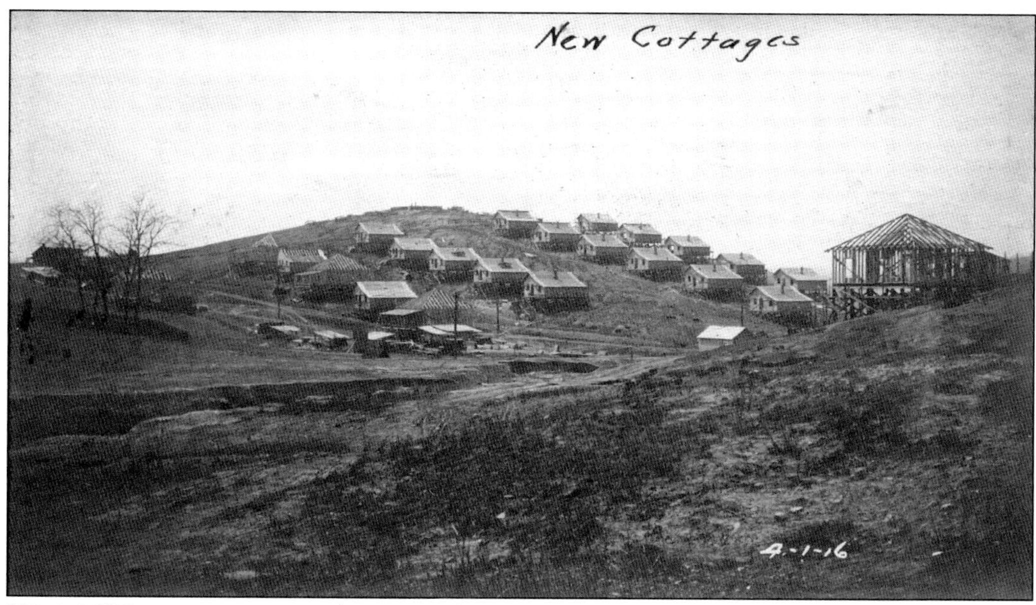

While TCC was constructing the No. 2 plant, the company was also preparing for employment expansion by building new worker housing adjacent to its existing housing at Newtown. This photograph of new cottages under construction on Cemetery Hill was taken in April 1916. The cemetery is visible in the distance, at the top of the hill.

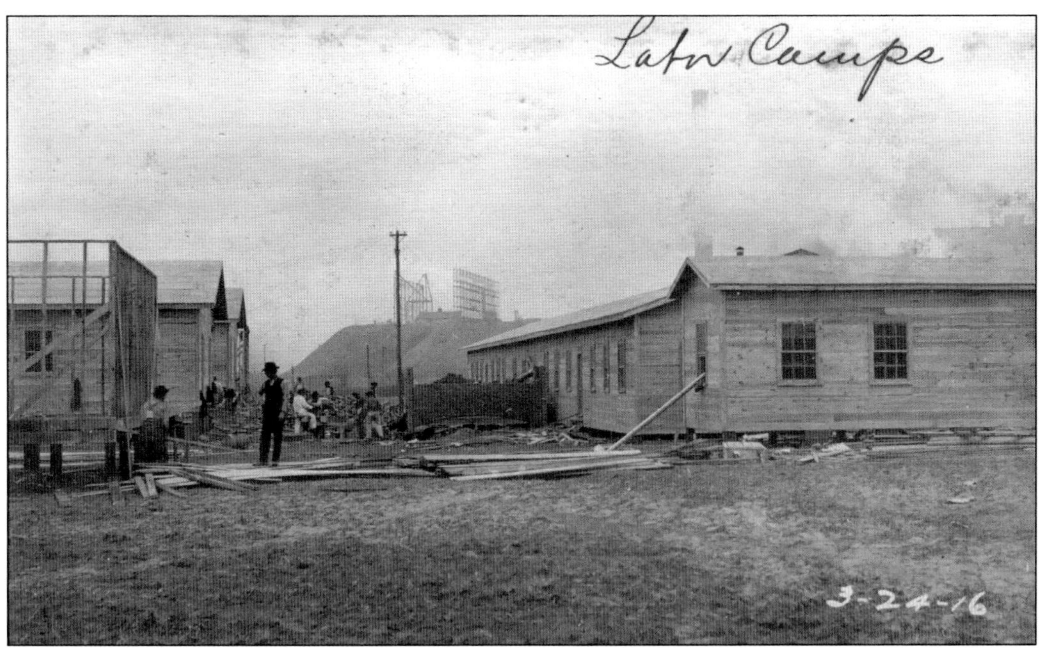

Due to the scarcity of labor, a semipermanent colony of foreign workers was established by TCC in 1906. Because most of the workers spoke no English, depending on foremen for translation on the job, the company created dormitory-style housing for them in a camp on the south side of the Ocoee River across from Isabella Junction. A footbridge led across the river to the plant site. In these photographs from March 1916, the foreign labor camps are being expanded to provide housing for additional workers, most of whom would be employed at one of the two acid plants. The skeleton of the No. 2 plant is visible in the background of both photographs. Its progress over the course of a week, and its location in relation to the No. 1 plant, can be seen below.

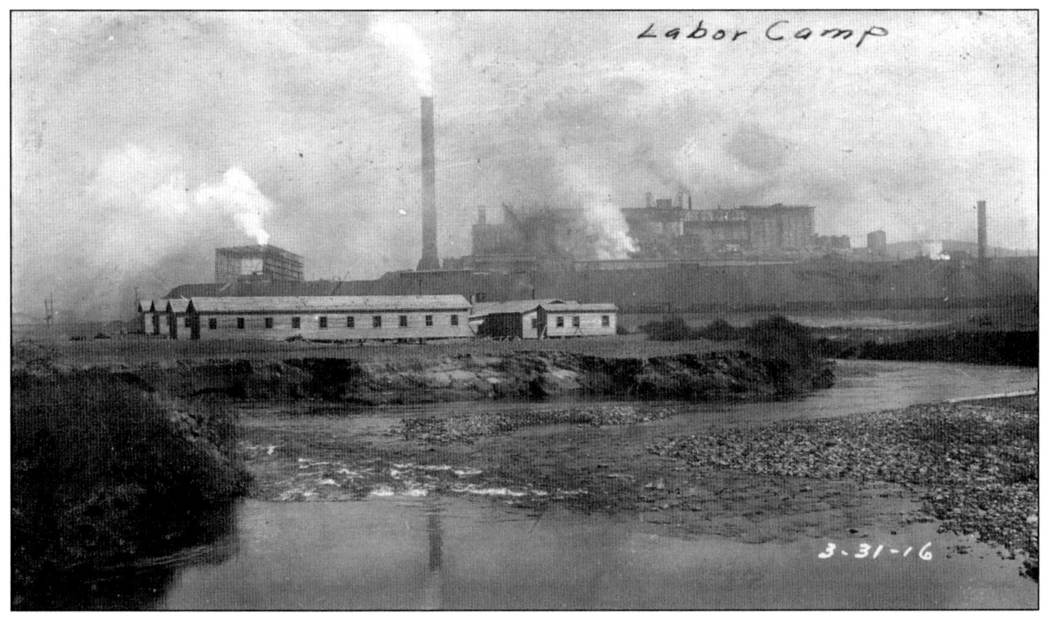

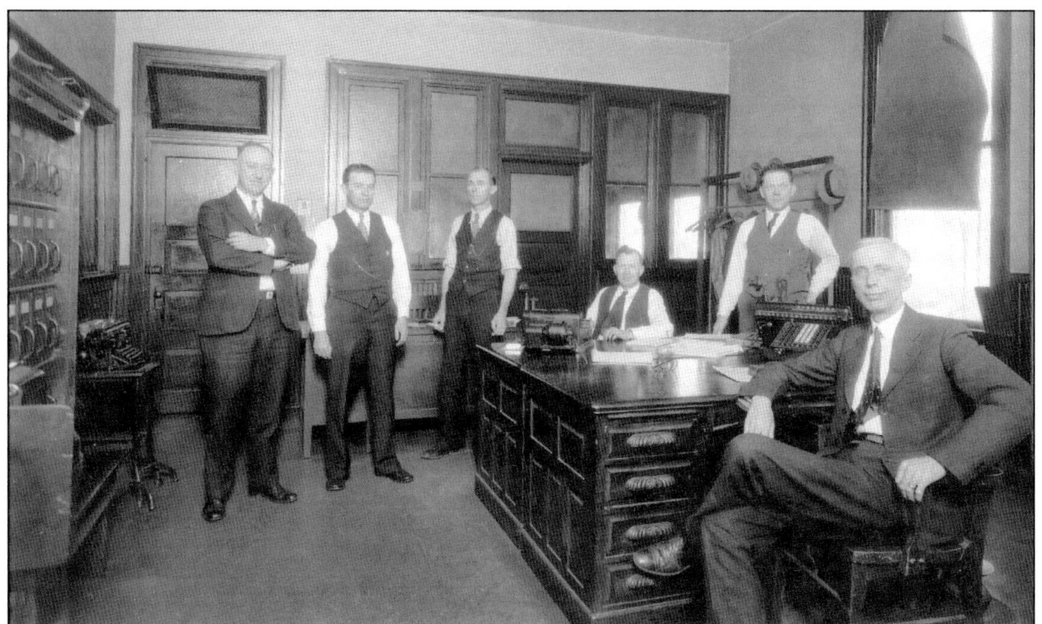

This photograph of the Tennessee Copper Company accounting staff was taken in the old general office building and probably dates from the 1930s. From left to right are chief clerk Paul Norton, Robert L. Frye, Julius Taylor, W.P. "Pat" Terry (who later served as mayor of Copperhill), C.E. Mock, and L.A. Denman. (Author's collection.)

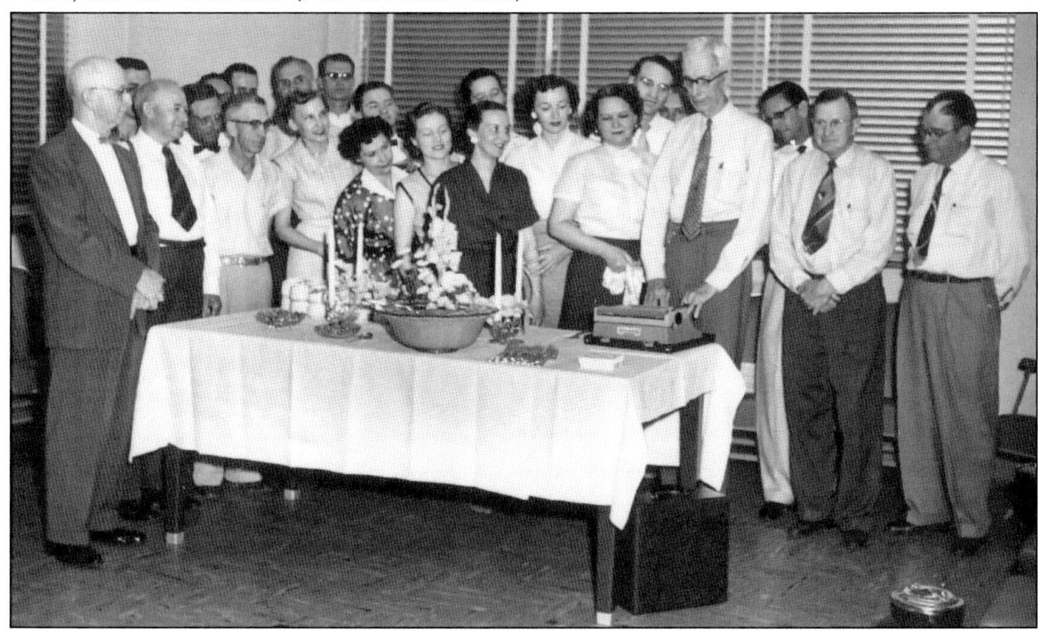

Pictured in this April 1955 retirement photograph are, from left to right, (first row) TCC chief clerk R.E. Barclay, Avery Parish, unidentified, Margaret Duncan, unidentified, Eva Lou Boling, Mary Claude Bidez, unidentified, retiring metallurgical accountant L.A. Denman, W.P. Terry, and Robert L. Frye; (second row) unidentified, W.H. Ritchie, two unidentified, Louis Ward, Oliver "Brud" Hawk, Cecil Jones, unidentified, Ann Mull, Otto Neal, and Wallace Whitfield. (Author's collection.)

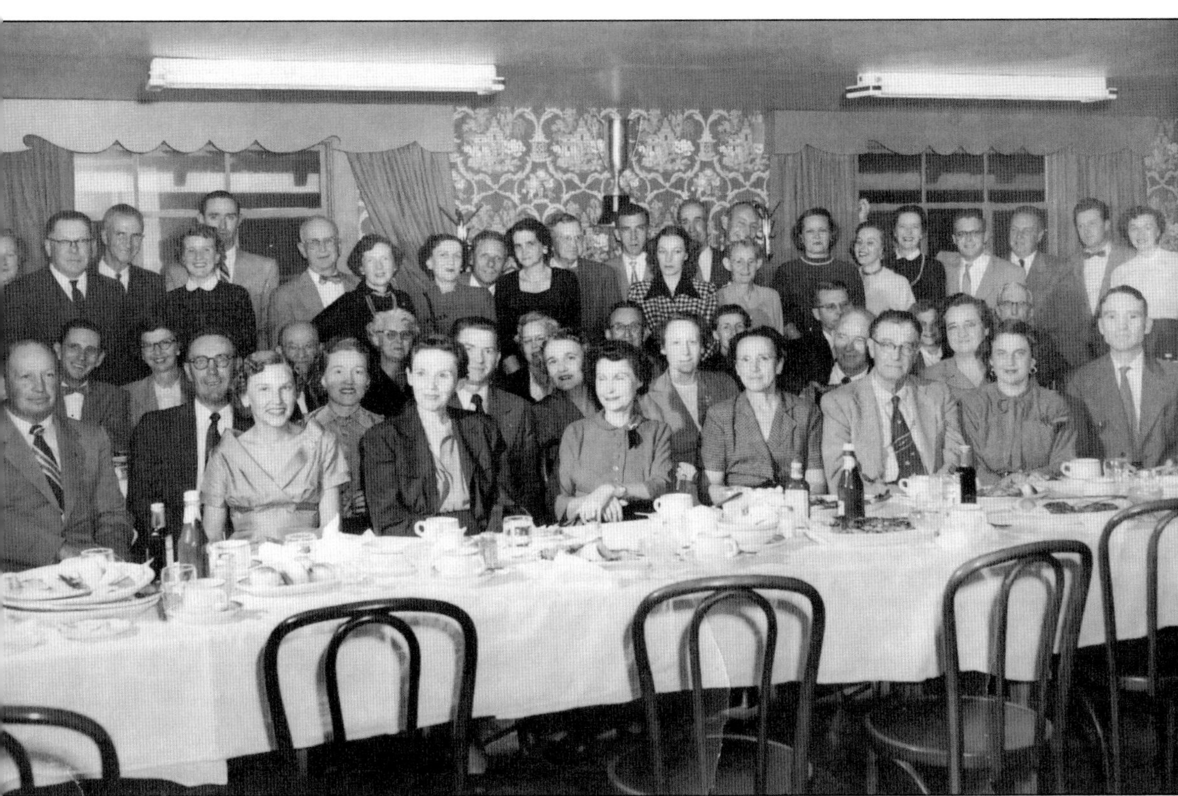

This photograph of the TCC general office staff, along with spouses and guests, was taken at a holiday dinner in December 1955 at a restaurant on Blue Ridge Lake. The diners are, from left to right, (seated at the table, first row) unidentified, Margaret Duncan, Grapelle Mock, unidentified, Gertie Frye, Robert L. Frye, Bea Neal, and Otto Neal; (seated, second row) Sam Boling, Eva Lou Boling, Cecil Jones, Jewell Jones, unidentified, Avery Parris, and Edna Weeks; (seated, third row) Wallace Whitfield, Geneva Whitfield, two unidentified, Ida Belle Earnest, Zellie Earnest, Louise Hawk, Oliver Hawk, Genevieve Denman, and L.A. Denman; (standing, fourth row) C.E. Mock, Jean Zachary, MaryLu Barclay, Lena Terry, Ethel Ritchie, unidentified, and Jottie Loudermilk; (standing, fifth row) Jessie Ward, Louis Ward, Dr. W.C. Zachary, R.E. Barclay, W.H. Ritchie, W.P. Terry, unidentified, Carl Loudermilk, two unidentified, Mary Claude Bidez, unidentified, Vernon Bidez, two unidentified, and Ann Mull. (Author's collection.)

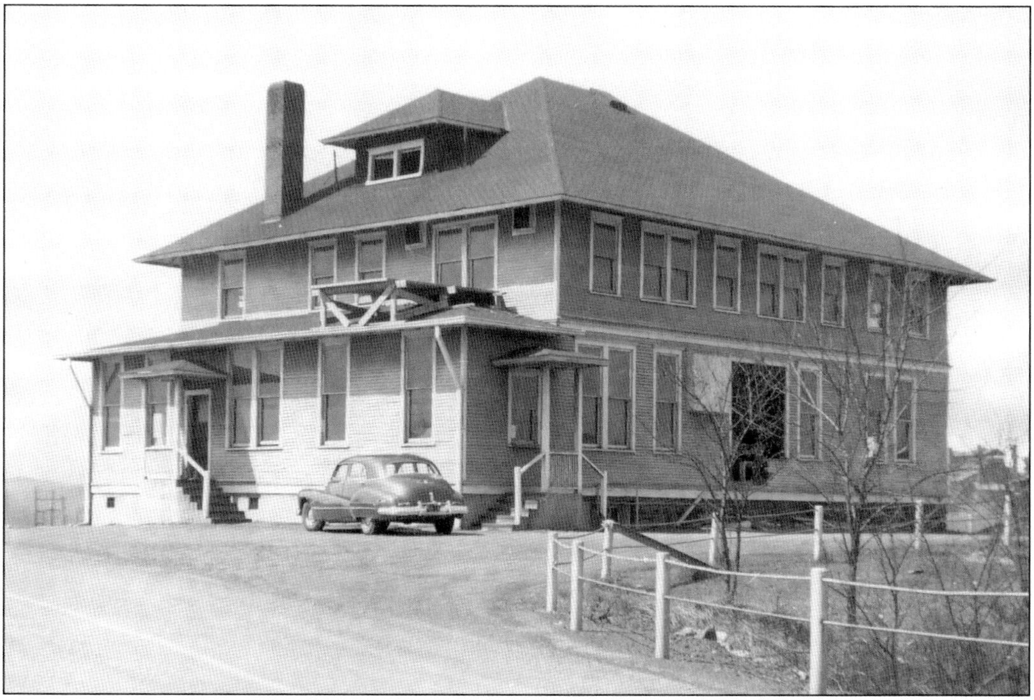

The image above is the last known photograph of the TCC general office before demolition began in the late 1940s to make way for a modern office building (below) at the same site. During the construction period, office personnel were moved temporarily to the Blue Goose hotel. The new offices were occupied in mid-1950 and are still standing.

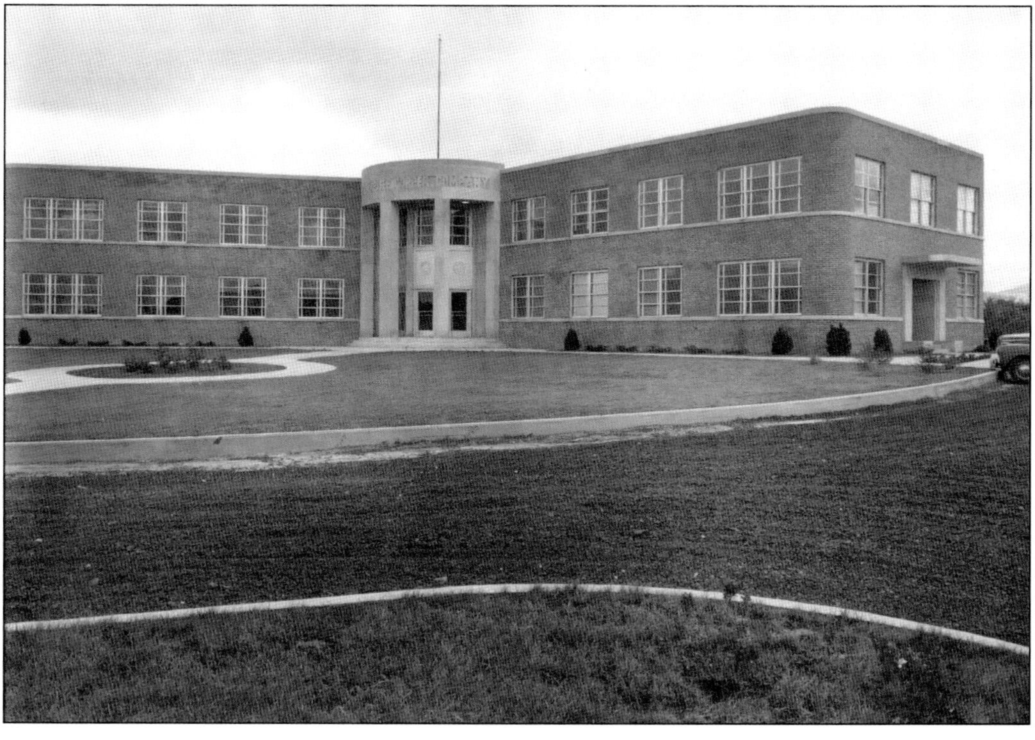

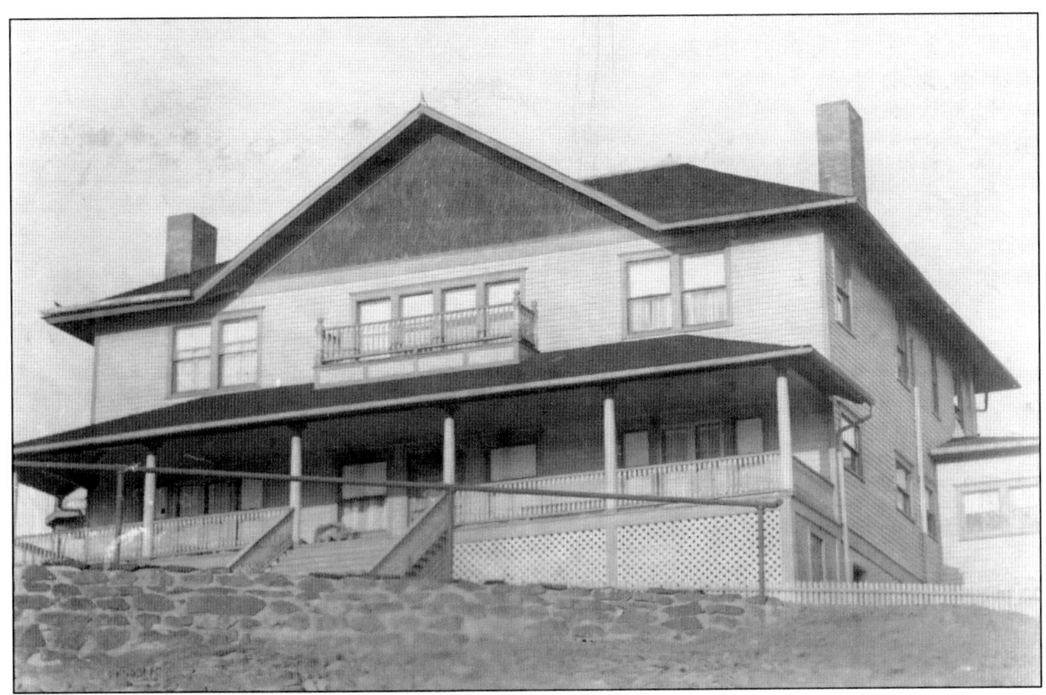

In 1909, future TCC president Utley Wedge made a visit to Copperhill and sent his wife a photographic postcard (above) of "the manager's house" where he had been staying. This appears to have been the original TCC general manager's house that can be seen in other early photographs. The general manager's residence remembered by Basin residents today was the later one, below, which was occupied for many years by TCC president and general manager T.A. Mitchell. In 1976, it was razed by TCC's successor, Cities Service Company, along with all the other remaining company housing on Smelter Hill.

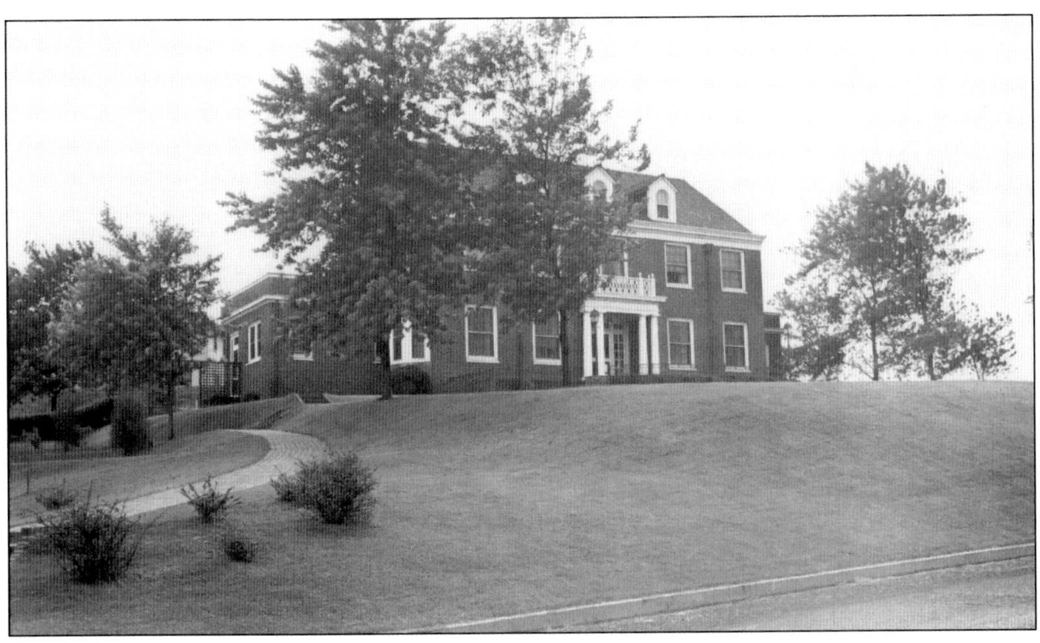

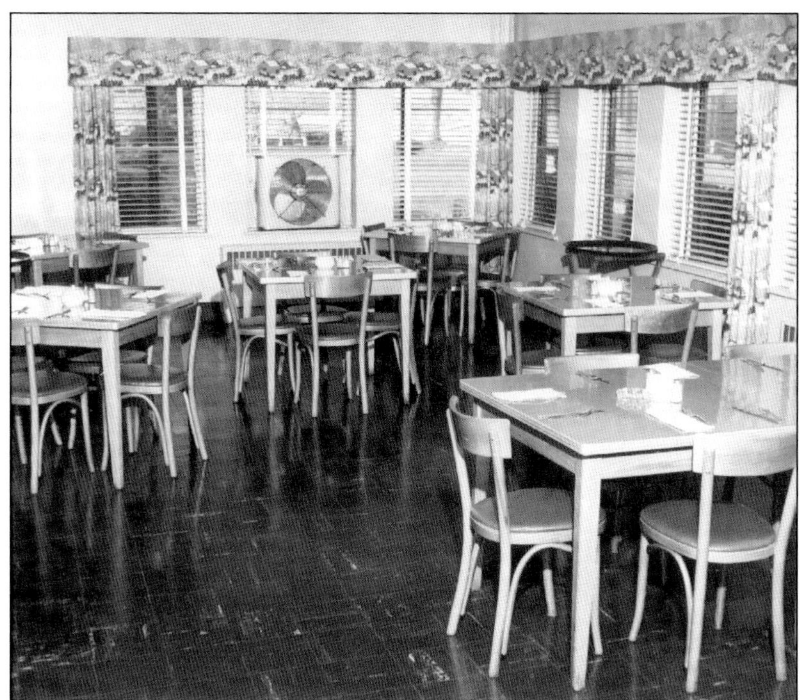

As TCC began modernizing its facilities on Smelter Hill in the late 1940s, one of the buildings that found itself back on the drawing boards was the Blue Goose residential hotel. The original building was replaced by a long, low two-story brick building with a landscaped lawn. This photograph of the dining room in the new building was taken in March 1960.

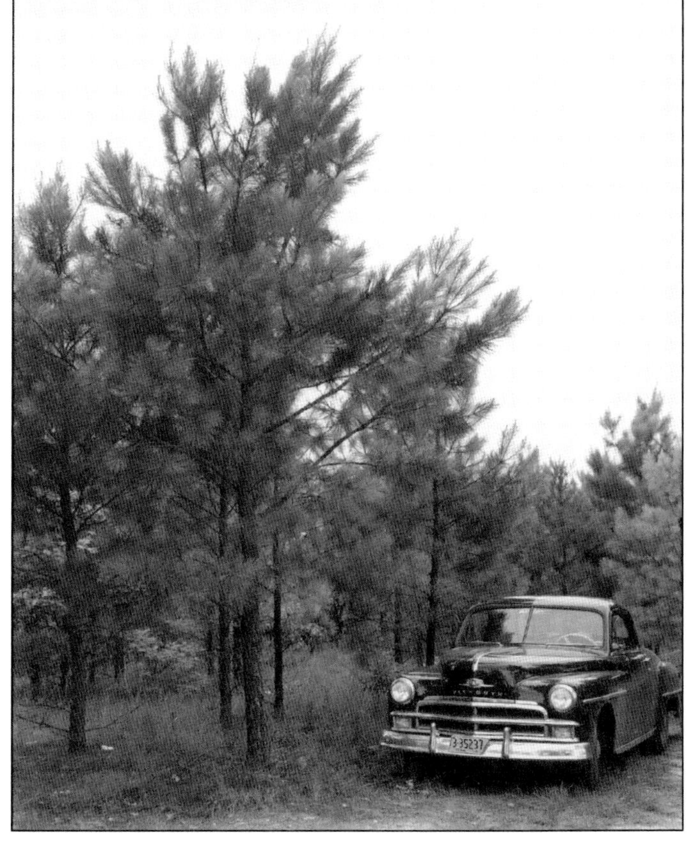

By the early 1940s, TCC had begun to focus on environmental reclamation and had planted the first pine trees on the bare red hills. This photograph from the files of E.H. Westlake shows the results after a few years of growth. In the early 1950s, the reclamation efforts were expanded to include the planting of a miraculous Japanese erosion-control plant called kudzu.

At the TCC mines, a system of check tags was used as a safety precaution to make sure all visitors who were not regular crew members were out of the mine before the regular blasting time at the end of the shift. In this June 1954 photograph taken at the Mary Mine headframe, safety engineer A.D. Annand is using the check tag system as he prepares to go underground.

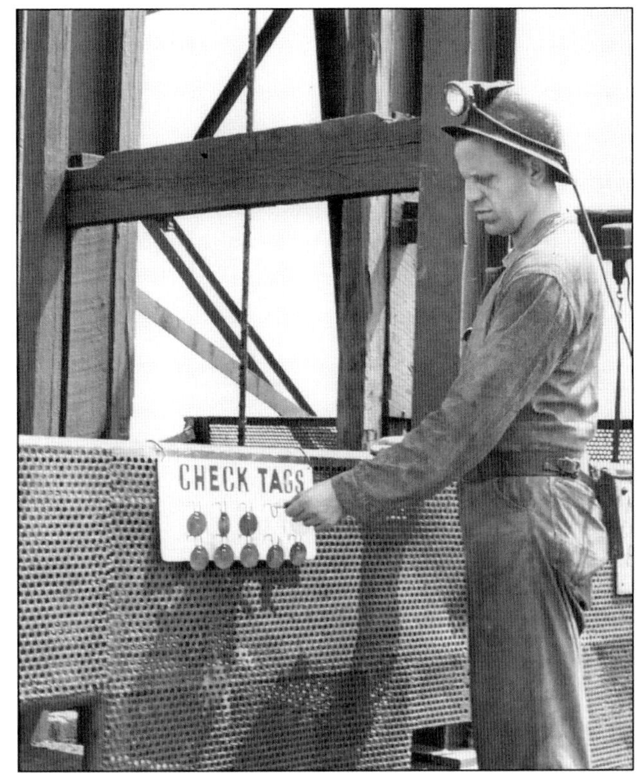

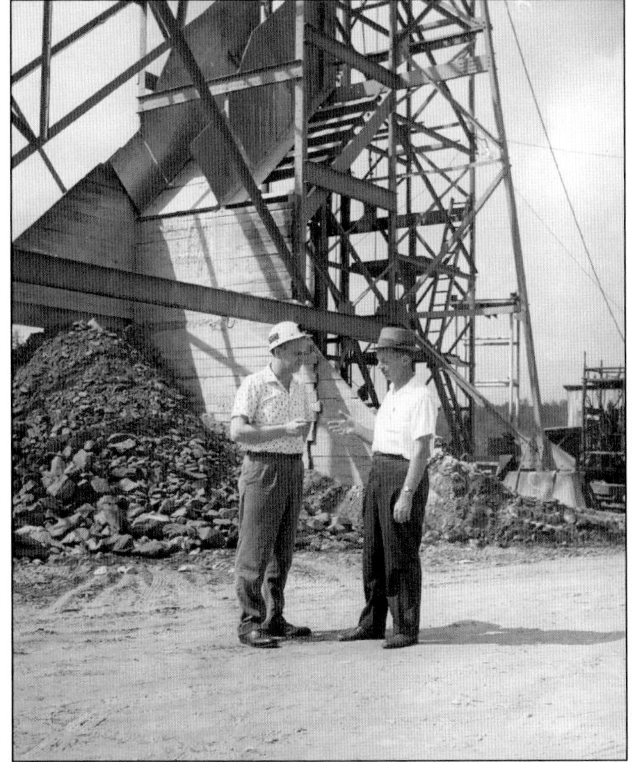

In 1952, when Lamar Weaver was promoted to general superintendent of TCC, H.F. Kendall succeeded him as superintendent of the mining department. His former coworkers remember that Kendall, who had joined the company as a geologist in 1928, was almost never without his trademark hat. In this 1960 photograph at Cherokee Shaft, Kendall is on the right, and safety engineer A.D. Annand is on the left.

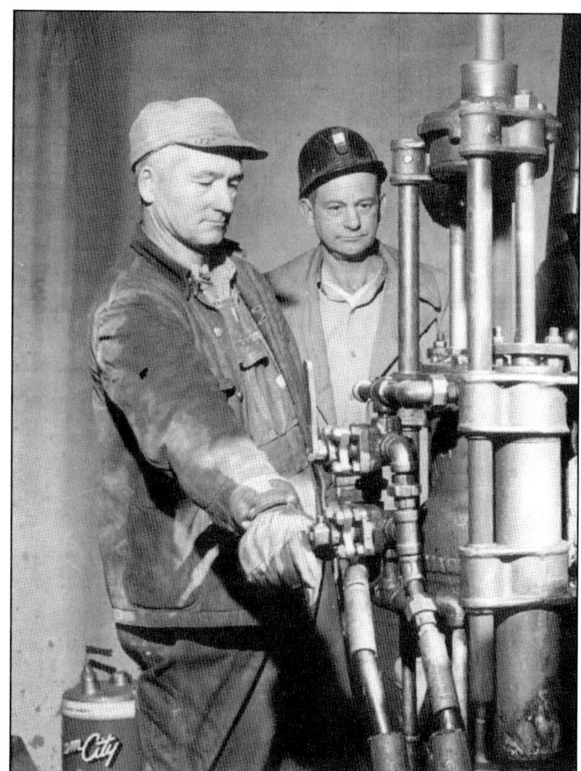

In 1956, TCC began the sinking of the B shaft at Calloway Mine. After the initial shaft was sunk, diamond drilling was used to explore the ore body near the lower levels. Diamond driller Cecil Bruce was at the machine when the drilling of a new pilot hole was started in spring 1956. The hole ultimately went down about 350 feet. Looking on is Calloway construction foreman E.H. Hawkins.

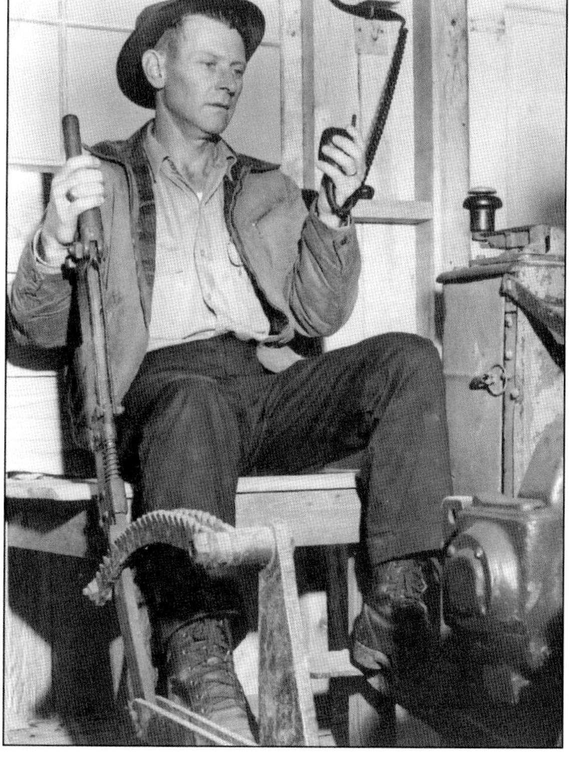

A cable going through the pilot hole created by the diamond drill at Calloway B shaft allowed for the lowering of a "birdcage," a drilling platform for two miners whose job was to enlarge the pilot hole to a larger shaft running all the way to the surface. In this November 1956 photograph, hoist operator Earnest German is in charge of raising and lowering the birdcage during construction.

In 1954, to supplement the company's regular on-the-job training, TCC initiated a home-study program for employees looking to improve their skills. Among the students to earn a certificate from International Correspondence Schools in 1958 was mine electrician Bazelle Nelson, who had been an apprentice at the beginning of his course. In this July 1958 photograph, Nelson prepares to go underground to make electrical repairs at Calloway Mine.

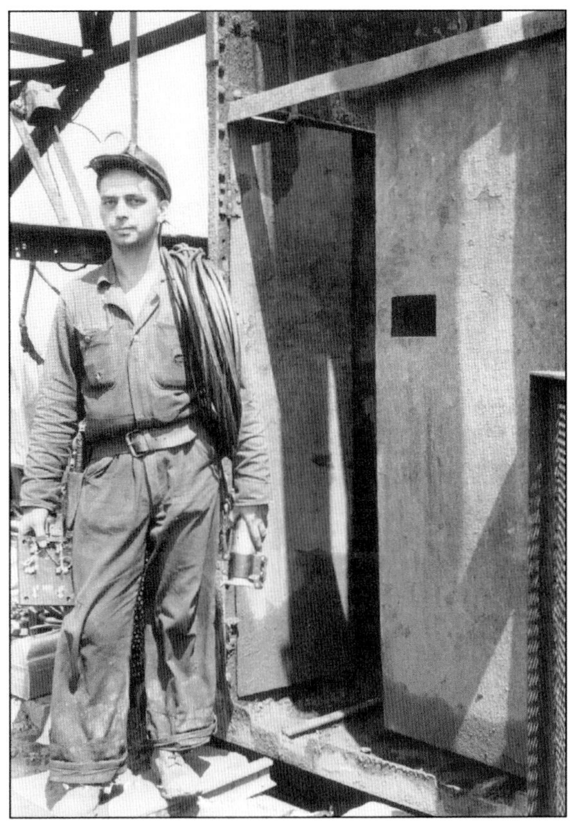

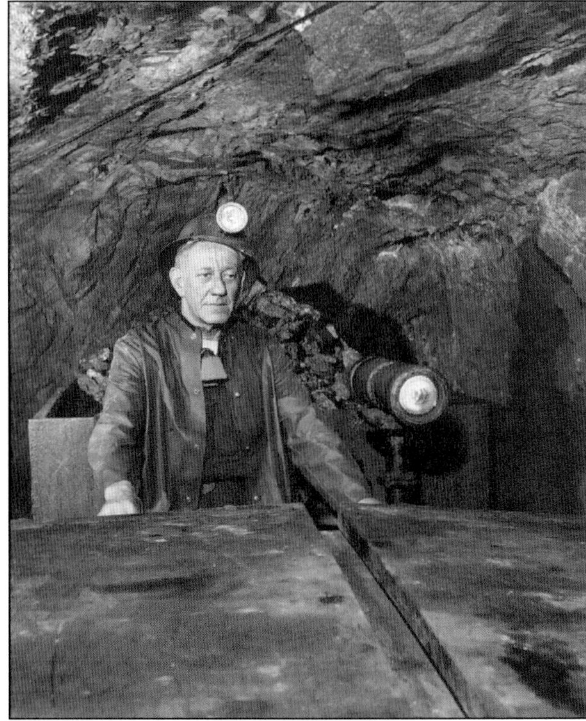

During work on Cherokee Drift in September 1960, motorman Luther Farmer waits near the face of the drift, ready to facilitate the filling and positioning of carloads of debris generated by the construction project. The filled cars were then hauled to Central Shaft, where the debris could be hoisted to the surface.

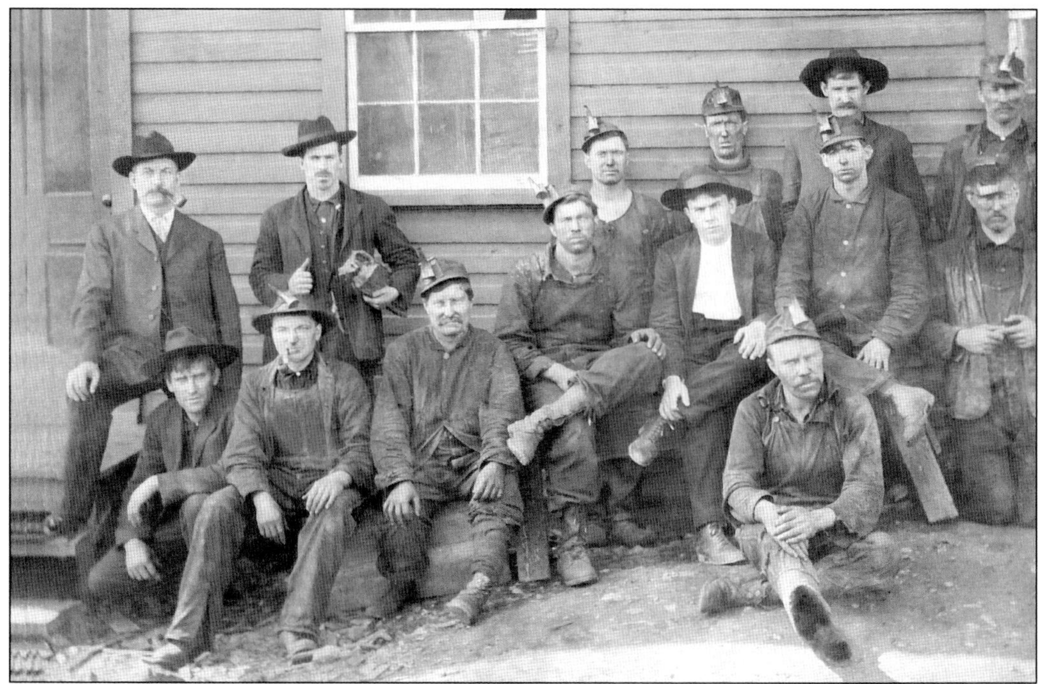

In the c. 1910 photograph above, miners in the Ducktown Basin are still wearing caps with open-flame oil-lit lamps. The lamps were officially called sunshine lamps but were known locally as "coon-shiners;" according to some older residents, the nickname originated from the miners' practice of economizing by using raccoon fat for fuel. Below, a group of teamsters and bosses from around the same time poses for a formal photograph.

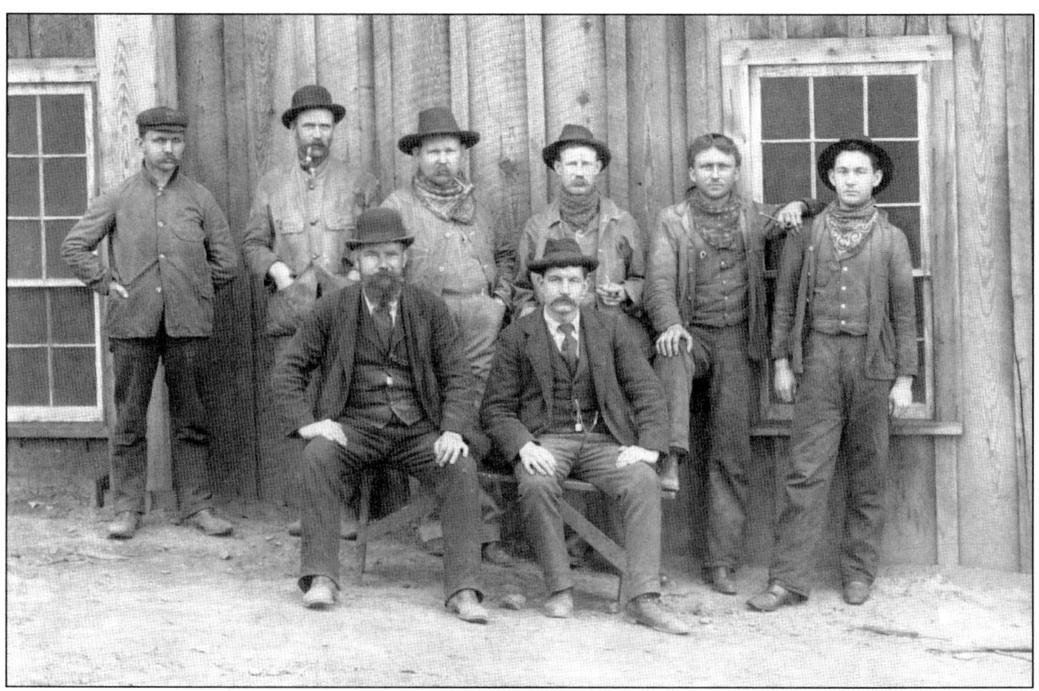

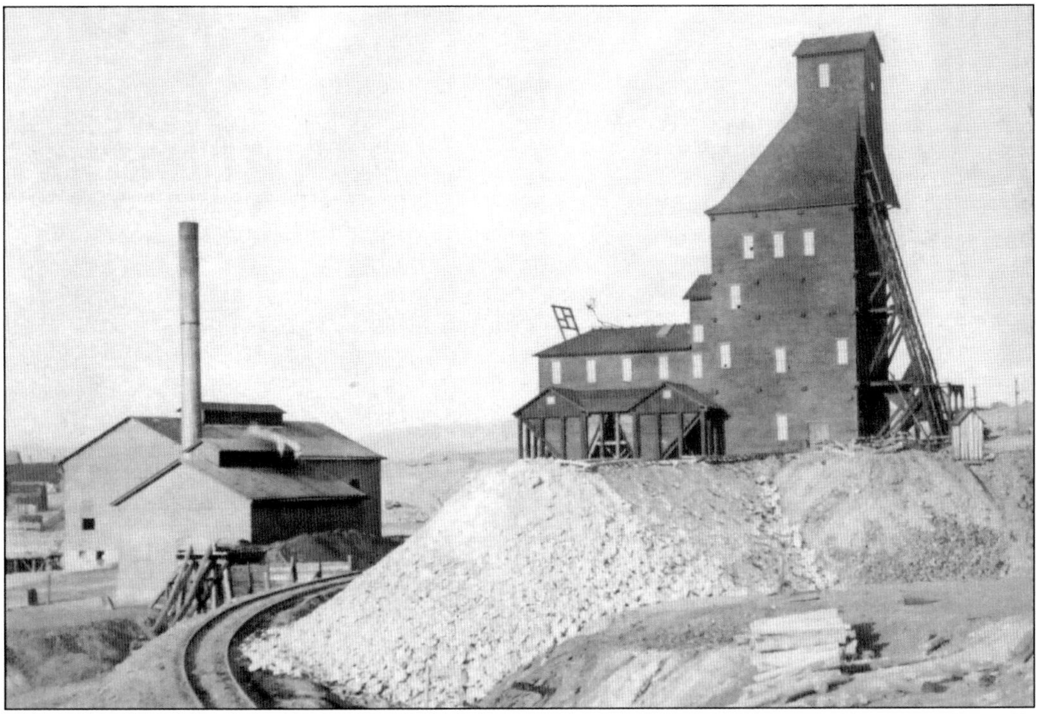

In 1899, TCC began work on the sinking of a new shaft at Burra Burra Mine. Above, in the early 1900s, the original headframe and ore bins overlook the hoist house and boiler house. Below, around 1945, the shop building stands on the hill to the left of the headframe and ore bins, which had been rebuilt in 1917, and the change house stands on the hill to the right. The hoist house is in the foreground at right. The railroad tracks visible in both photographs connected Burra Burra with the other TCC operations.

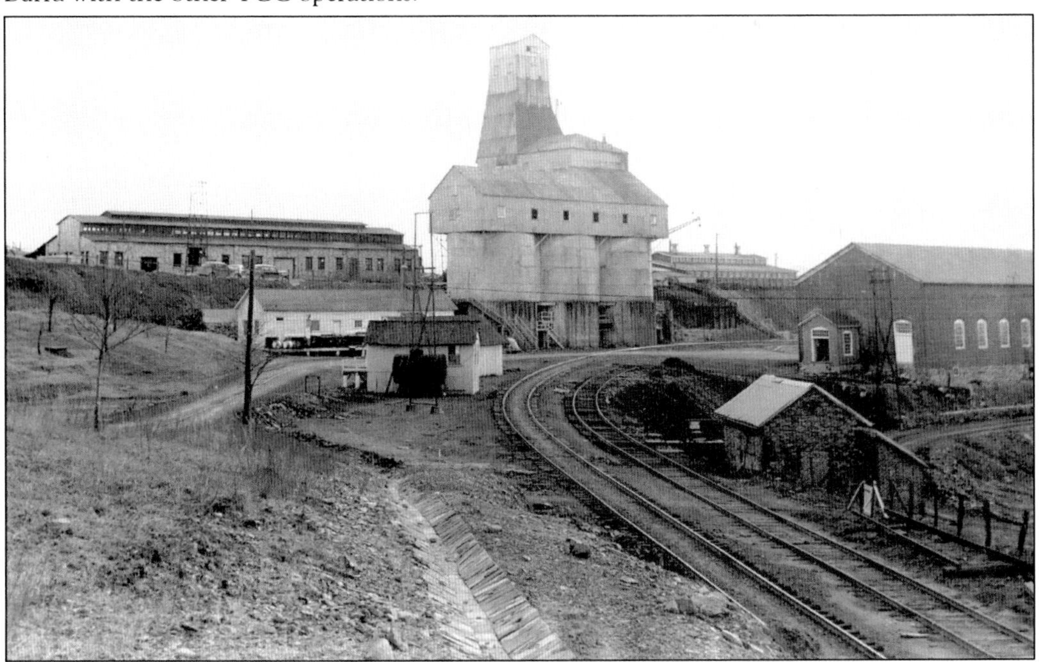

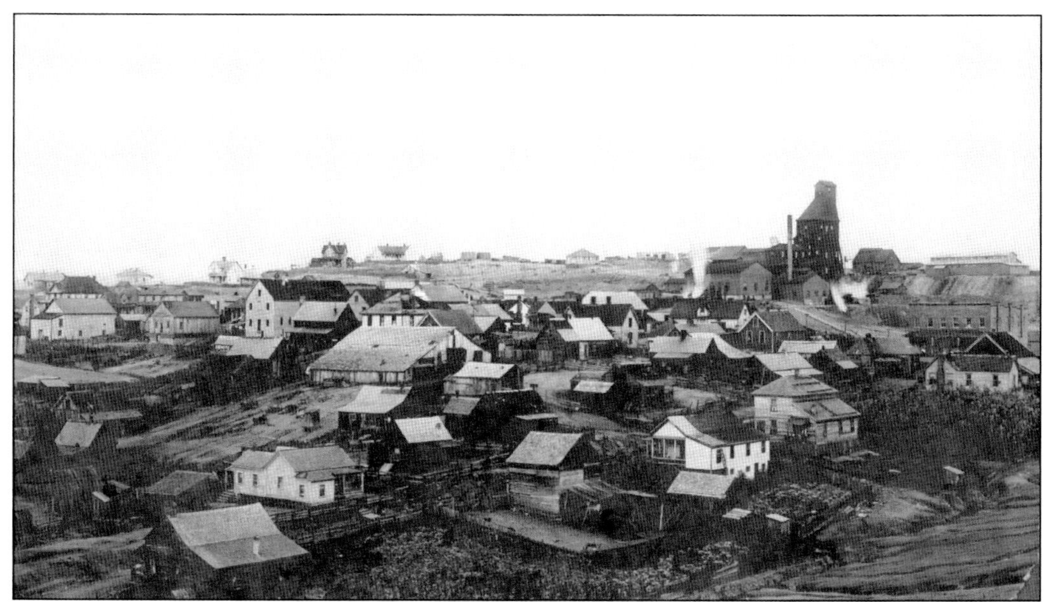

In the c. 1917 photograph above, the original Burra Burra headframe rises above the town of Ducktown, with mine buildings and managerial housing running along the ridge to its left. Below, in another early 1900s photograph, mine captain Charles V. "Cap" Skullman, fifth from left, and other staff members pose at the Burra Burra mine office. According to a 1914 account in the *Polk County Republican*, Skullman's hands were on the shovel when the first dirt was removed from TCC's Burra Burra shaft in 1899.

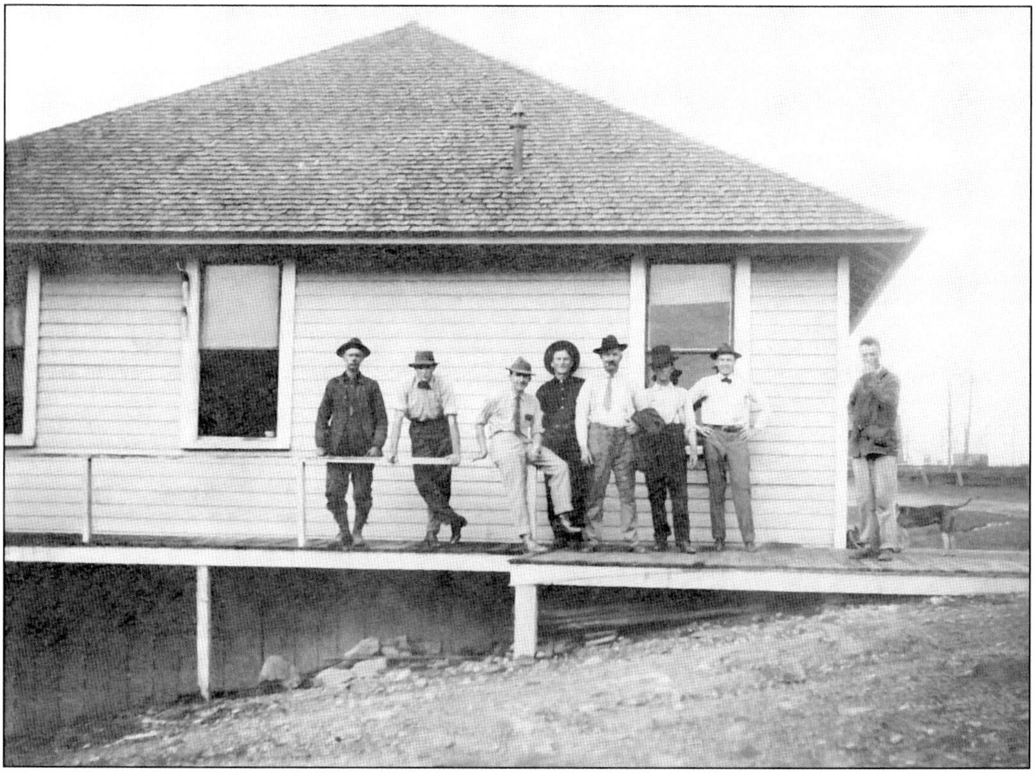

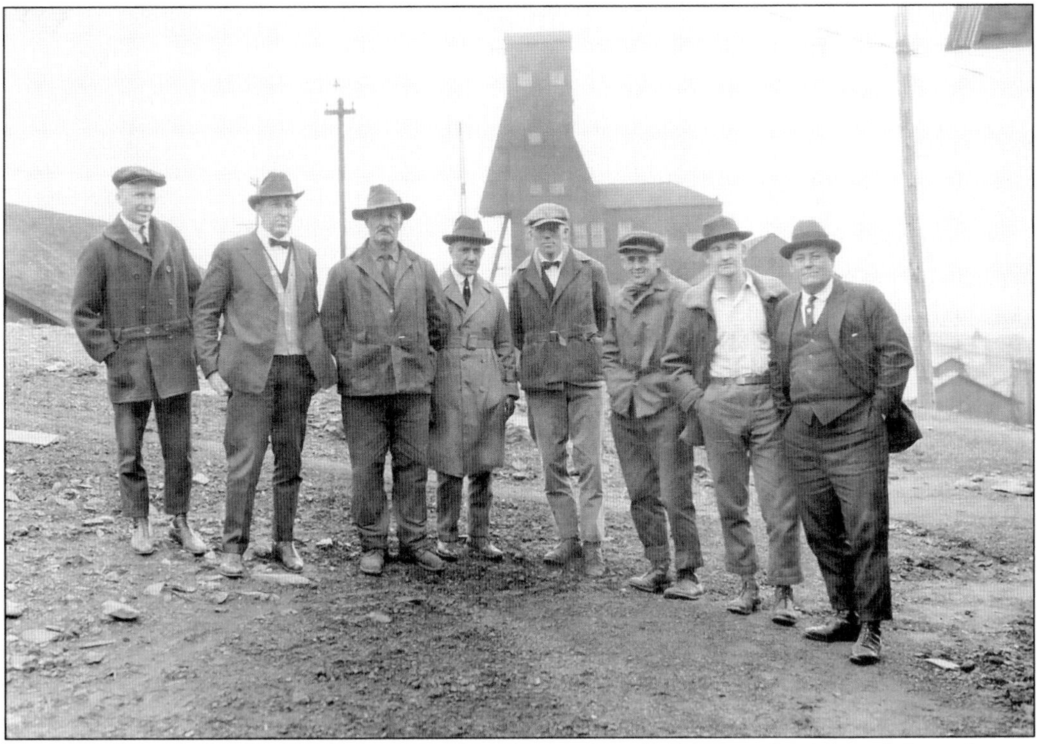

In this c. 1920 photograph outside the mine office, Cap Skullman is third from left, and mine superintendent H.B. Henegar is fifth from left. Sixth and seventh from left are two mining engineers identified only as McDonald and Johnson.

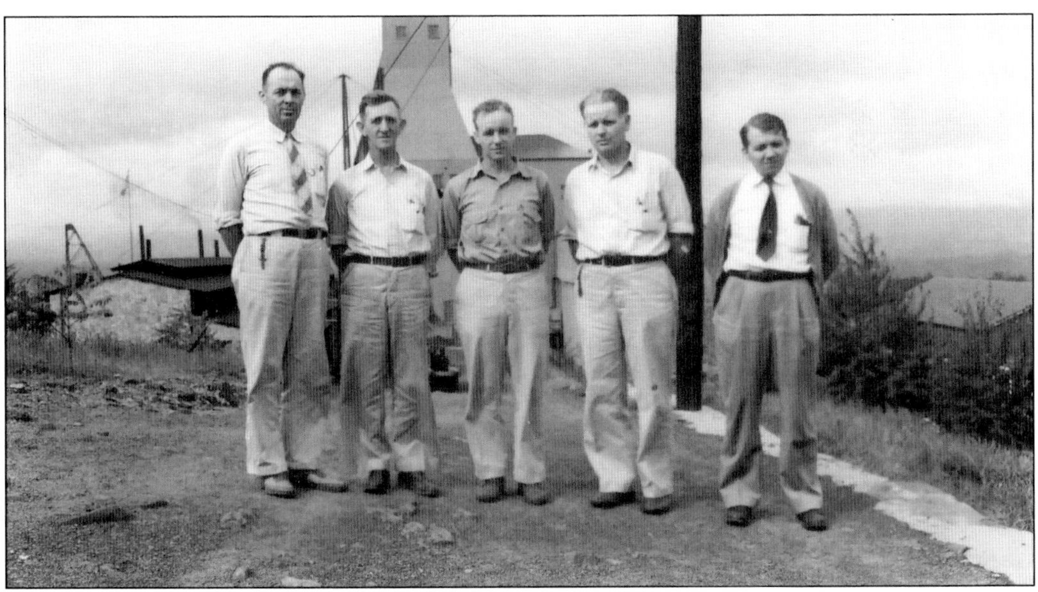

From left to right in this 1943 photograph are mine superintendent Lamar Weaver, safety engineer S.E. Sharp, geologist J.H. Ffolliott, safety engineer C.F. Seaman, and R.L. Beatty of the US Bureau of Mines. The Burra Burra shop building, constructed in 1922, can be seen on the left. (Courtesy of Alice-Ann Seaman Ferderber.)

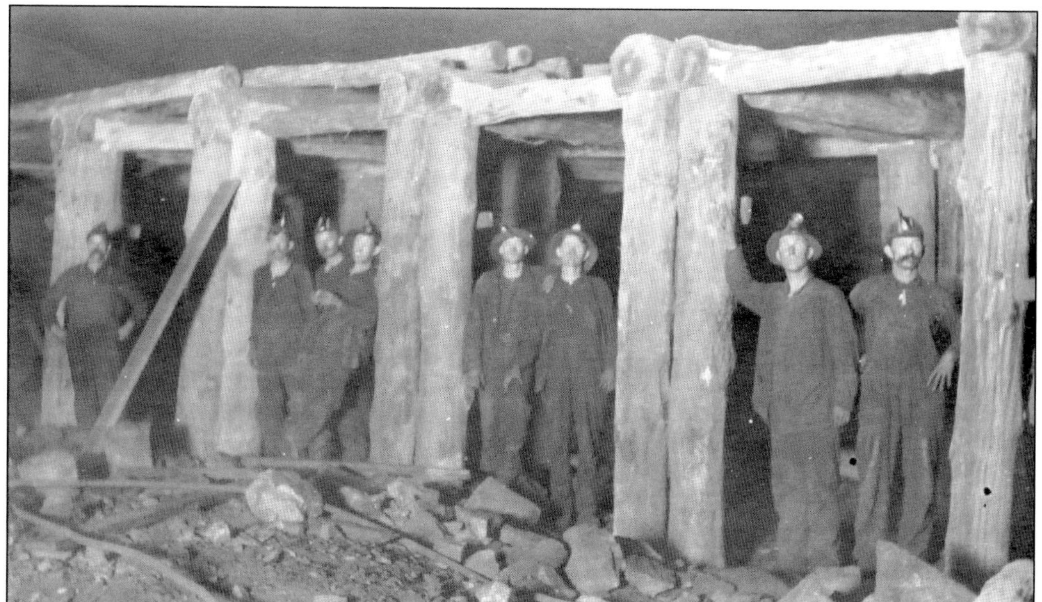

In this c. 1910 photograph, miners underground at Burra Burra stand in a timbered drift on the fourth level. On most of their caps are the open-flame coon-shiner lamps that were used before the advent of carbide lamps at TCC between 1910 and 1912. The man standing second from right appears to be wearing a carbide lamp.

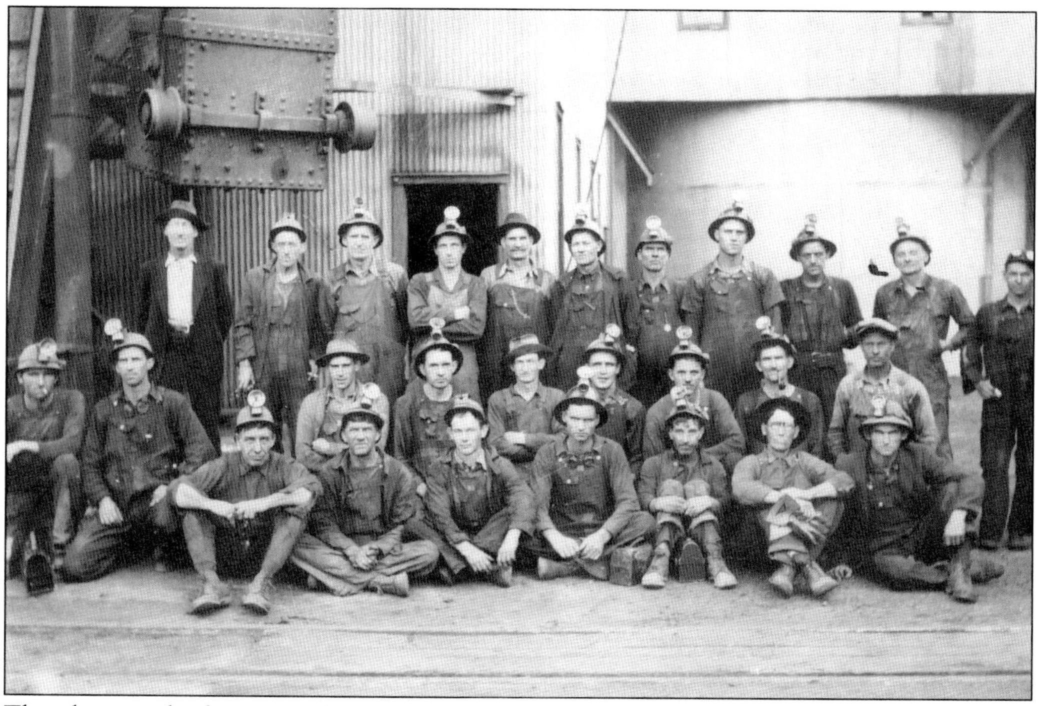

This photograph of miners and other crew members at Burra Burra was probably taken in the late 1930s. The miners' protective helmets have carbide headlamps, which would have been replaced in 1939 by electric headlamps. The only person positively identified in the photograph is Wesley Davis, who is seated second from right in the first row. (Courtesy of Peggy Walter Kilpatrick.)

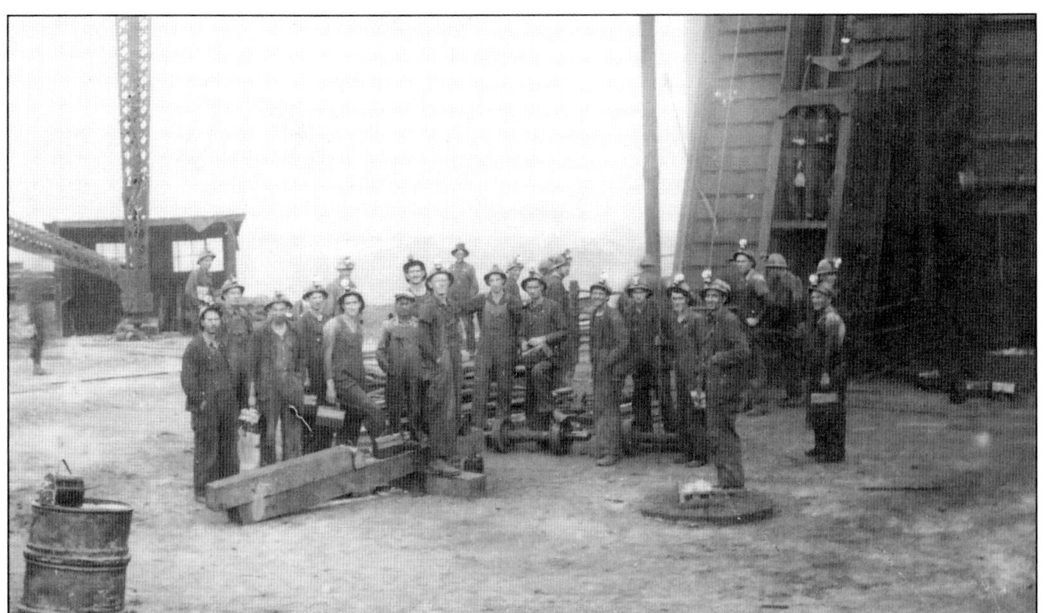

This photograph of miners on a lunch break at Burra Burra probably dates from the 1930s. At right, on the hoist that lowered workers into the mine, two men can be seen on the second level of the man cage. (Courtesy of Peggy Walter Kilpatrick.)

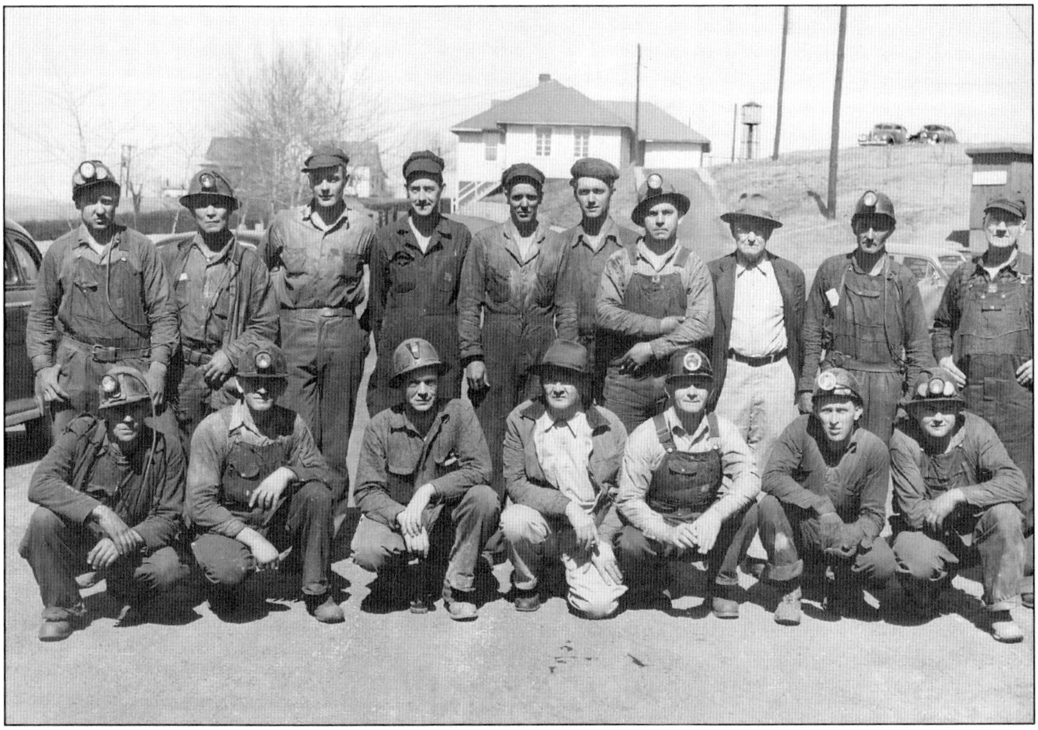

Third from left in the first row of this early 1950s photograph at Burra Burra is H.F. Kendall, who became superintendent of the TCC mining department in 1952. Fourth from left is ? Amburn, fifth from left is Clifford Helton, and sixth from left is J.R. Ramsey. In the second row, Wayne Mashburn is third from left, Kenneth Hughes is sixth from left, and Sam Craig is eighth from left.

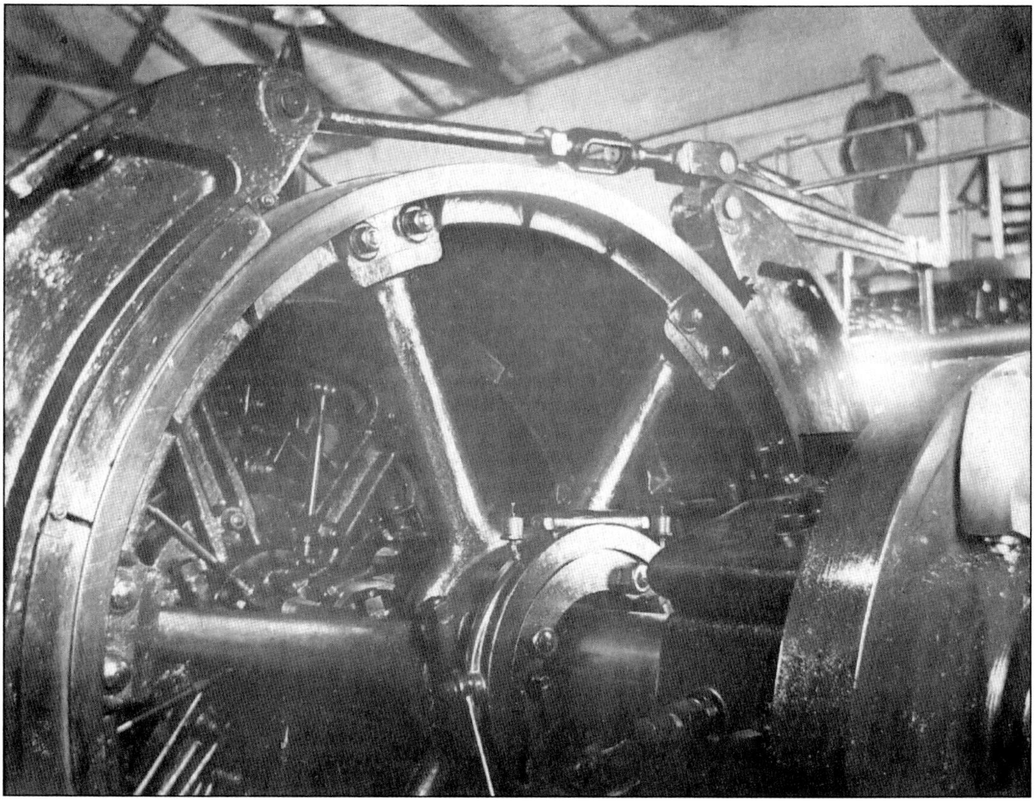

One of the earliest recorded images of TCC mining operations at Burra Burra was this dramatic photograph (above) of the hoisting engine, taken between 1900 and 1905 by Edward Oscar Boak. Below, a little more than a half-century later, the day and evening shifts at Burra Burra pose for a final crew photograph on May 28, 1958. The last ore was actually hoisted from the mine two weeks later, on June 13.

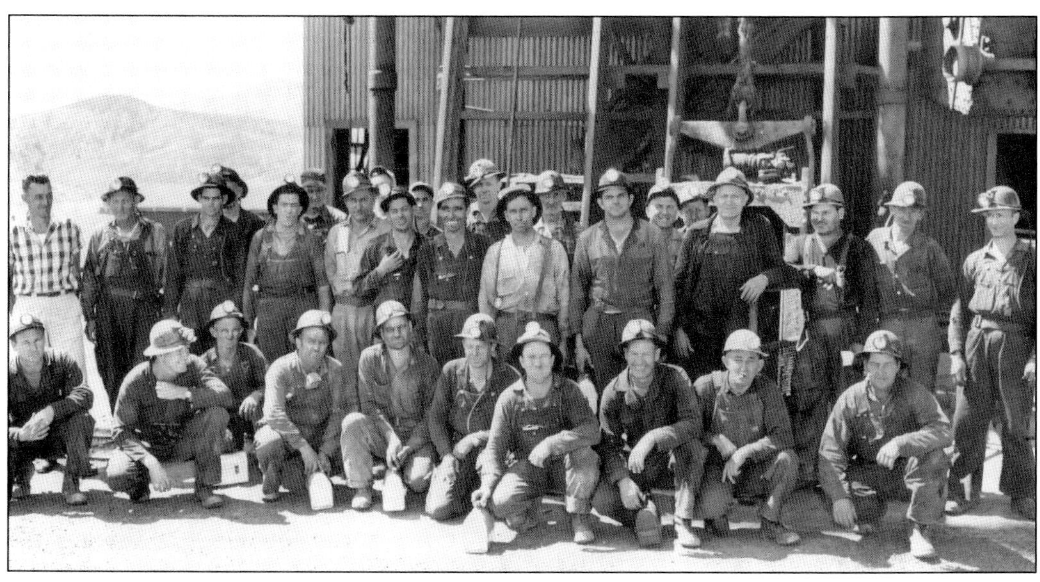

Two

City Life

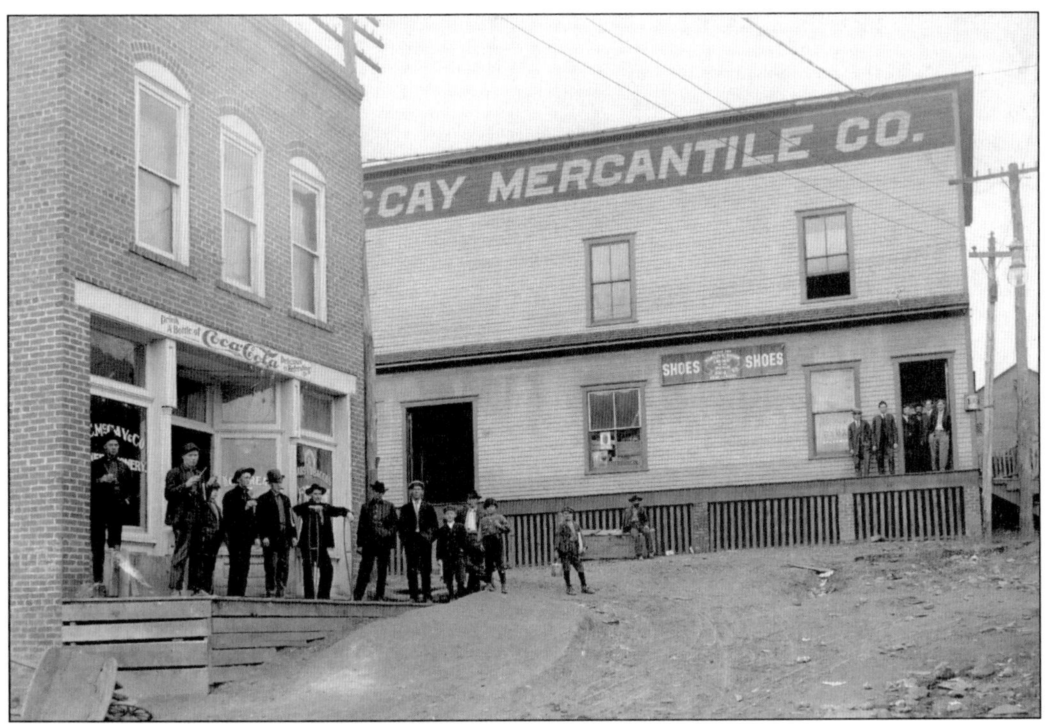

The Polk County Bank ("Capital $15,000.00," according to the window sign) is at the far right of this 1906 photograph of the McCay Mercantile Company, which stood on the site of the present Copperhill City Hall. D.C. McCay & Company, in the foreground, advertised confectionery, ice cream, cigars, tobacco, and 5¢ Coca Colas. Until 1908, when its name was changed to Copperhill, the town itself was also called McCays after its founder, Harbert T. McCay.

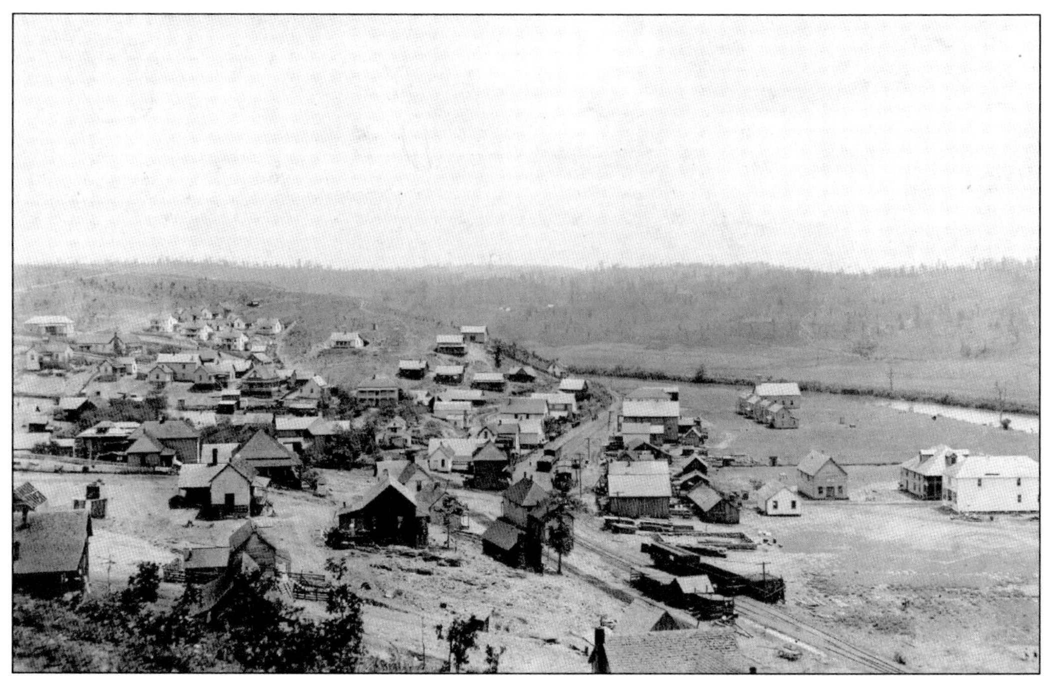

In 1904, the business district of McCays ranged along both sides of the railroad tracks, with very little development nearer the river. One exception was the original First Baptist Church in present-day McCaysville, Georgia, which can be seen at center right just beyond the four small houses. A loaded ore car sits in front of the McCay Mercantile building, at the bend of the railroad tracks.

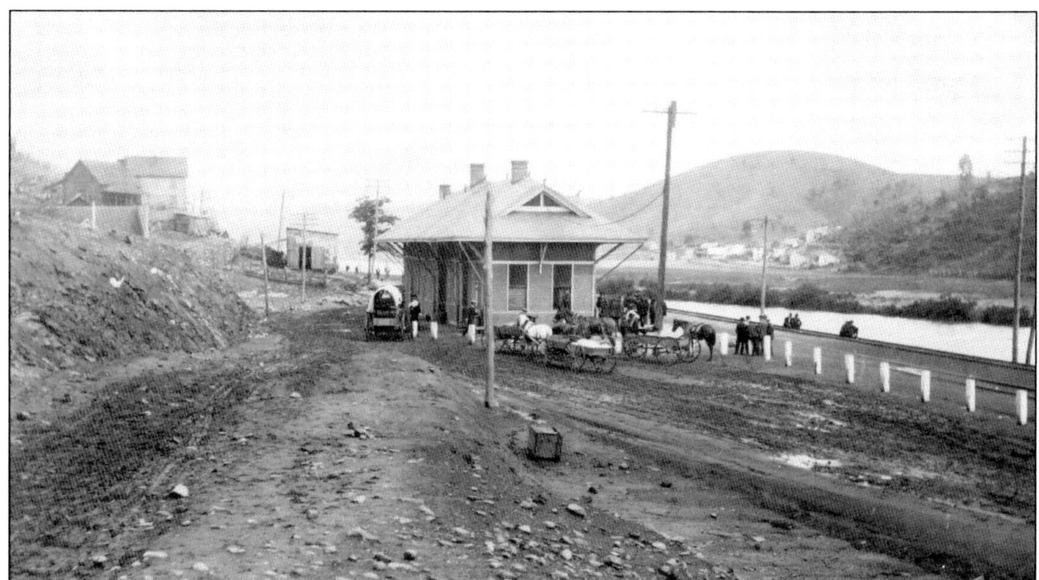

In 1907, the Louisville & Nashville Railroad began construction of new facilities just south of the TCC plant at McCays. In 1908, at the request of L&N and TCC, the town's name was changed to Copperhill, and the passenger depot was designated the Copperhill station. The McCays Post Office also changed its name to Copperhill, and the former Copperhill Post Office at Mary Mine became the Hyatt Post Office.

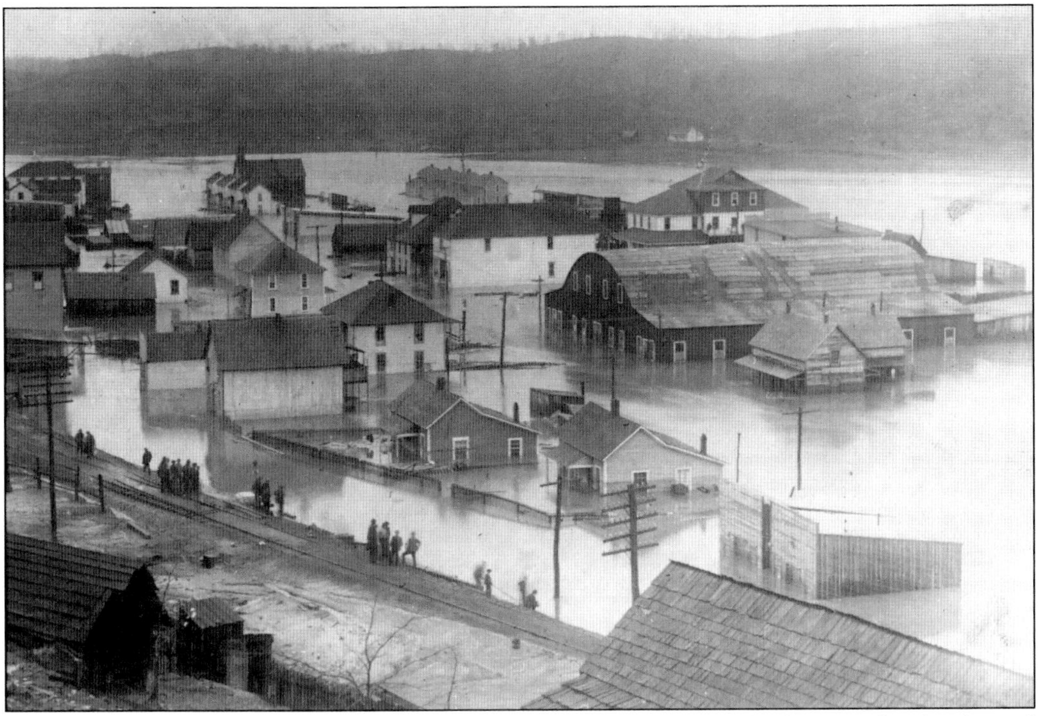

On November 18, 1906, after several days of rain, the Ocoee River broke out of its banks and raged through the lower part of the town of McCays. The water was reportedly as much as six feet deep inside some of the lower-lying buildings. In the photograph below, several pigs, who would have been allowed to run loose in town during this period, can be seen foraging at the edge of the water. Three people were drowned in the flood: a farmer named Cas Headrick, his five-year-old son Claud, and an unidentified black servant from the Randolph Hotel. Claud Headrick's small tombstone still stands in the Copperhill cemetery on Cemetery Hill. Both photographs were taken by E.H. Westlake.

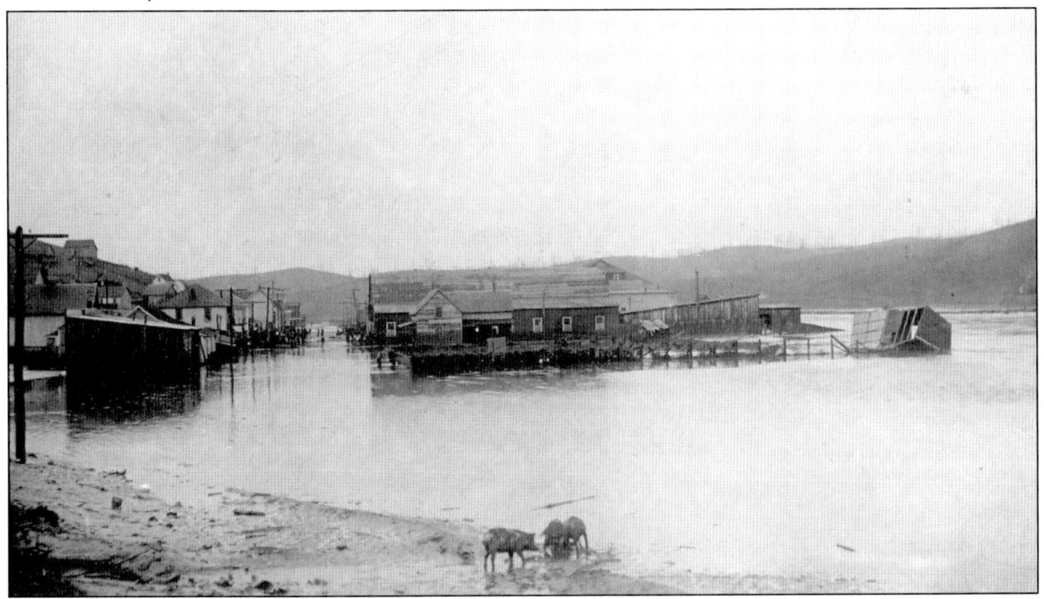

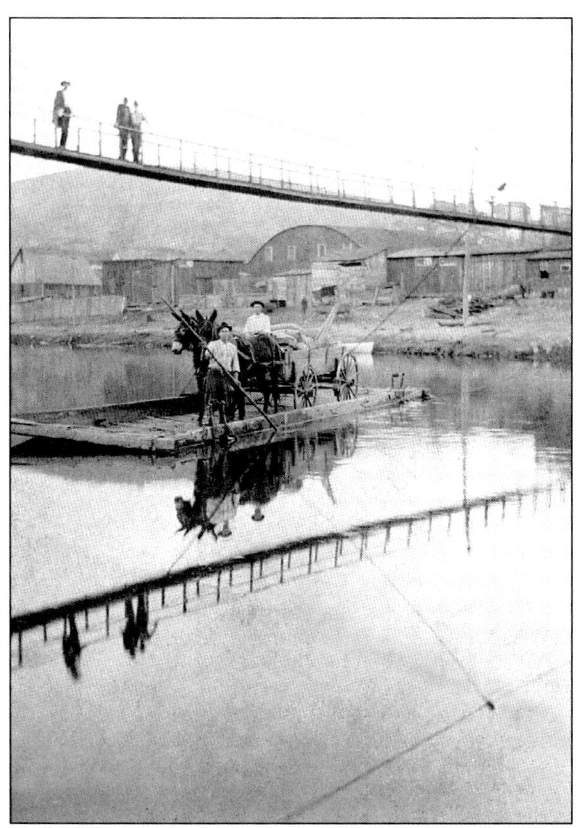

In this 1907 photograph, a mule-drawn wagon is ferried across the Ocoee River between McCays (later Copperhill), Tennessee, and Hendrickstown (later McCaysville), Georgia. The footbridge had been built in 1903 to accommodate pedestrians who preferred not to wait for the ferry. The oval roof of the skating rink can be seen in the distance on the McCays side of the river. (Courtesy of Greg Crawford.)

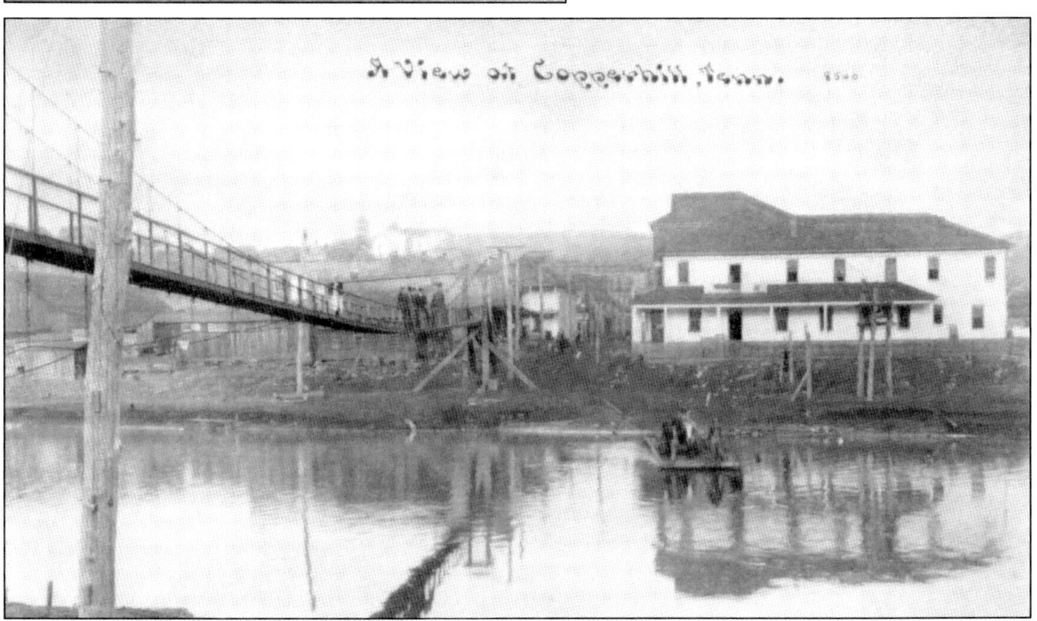

This postcard photograph was apparently taken between 1908, when the town of McCays was renamed Copperhill, and December 1910, when the Randolph Hotel (at right) on the Copperhill side of the river was destroyed by fire. In 1911, the footbridge was replaced by a steel bridge that could be used by both wheeled vehicles and pedestrians. (Author's collection.)

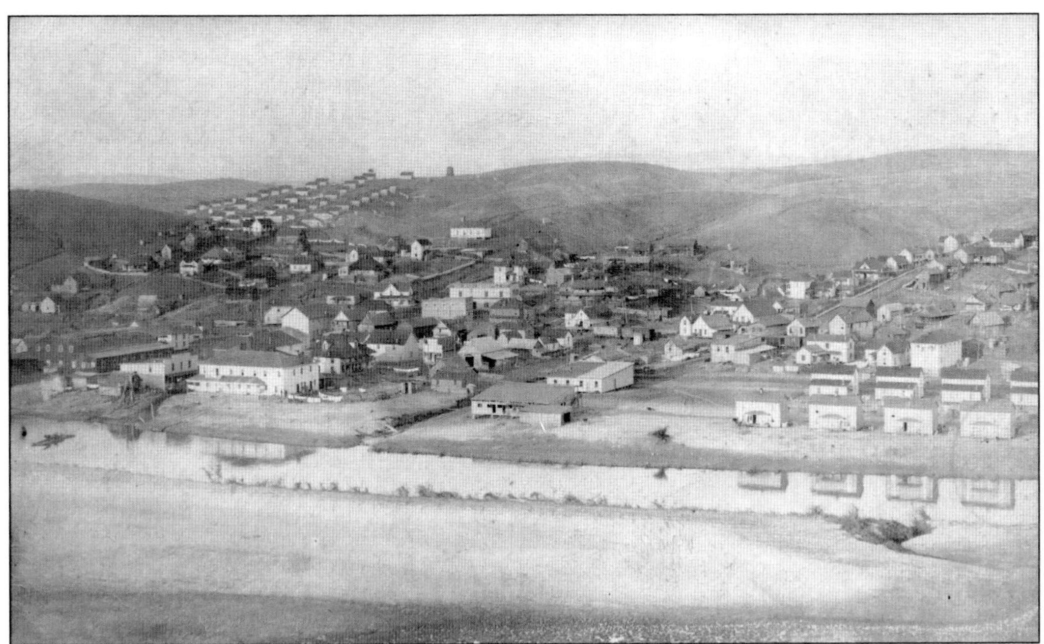

The Randolph Hotel is the two-story building in the left foreground of the image above. The creek near left center, which now runs underneath Ocoee Street, marks the site of the present-day concrete bridge, and the ferry is visible on the Ocoee River at far left. Below, a picture of a mass baptism in the river shows a more detailed view of the hotel and its surroundings. Both photographs were taken before the fire of 1910, which destroyed the majority of Copperhill's downtown business district.

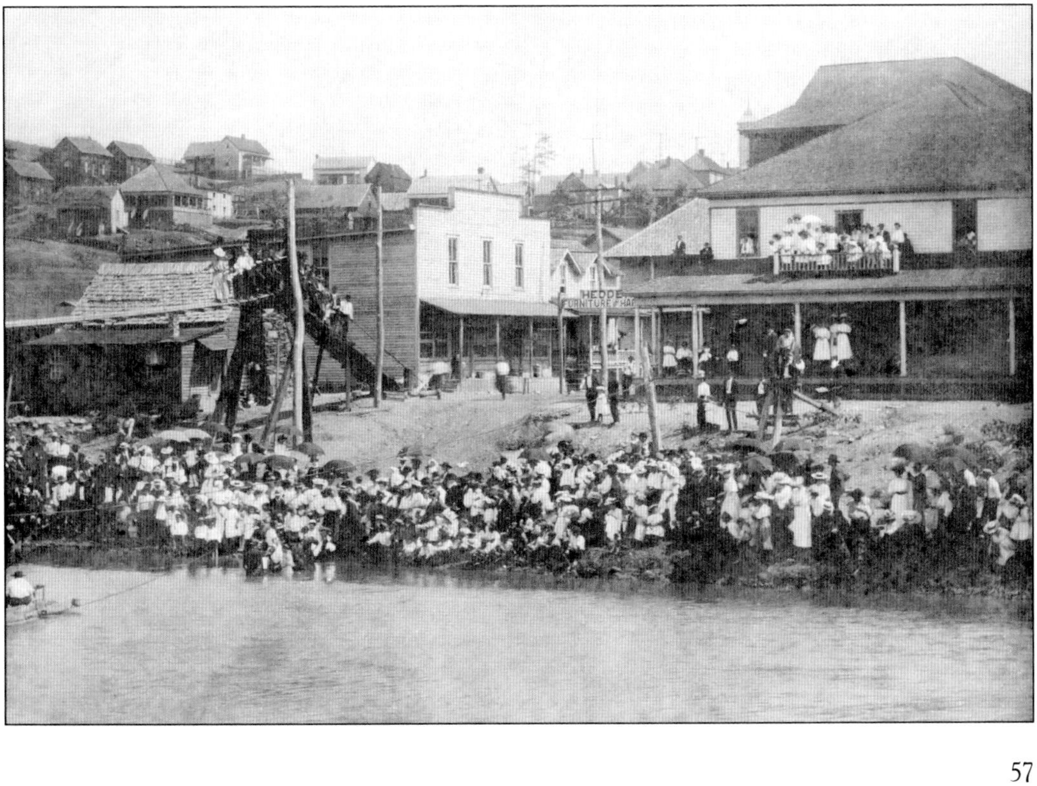

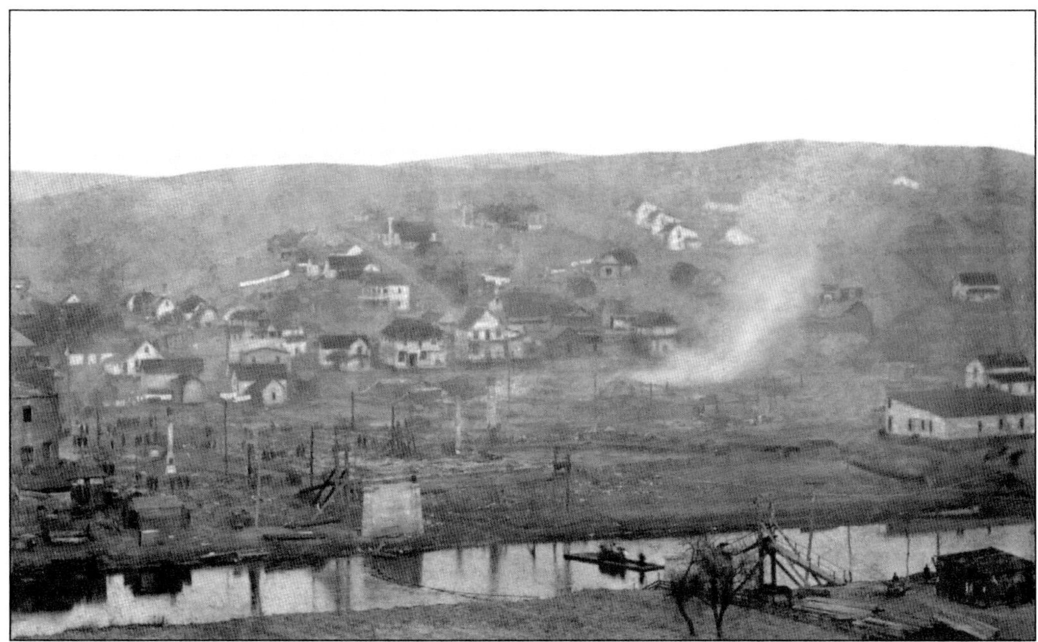

On the night of December 2, 1910, the city of Copperhill was ravaged by a fire that broke out near the Randolph Hotel, destroying the hotel, the footbridge, and more than 30 wooden buildings in its path. When morning came, much of the downtown business district was a smoldering ruin. Small figures can be seen at the left of this photograph, as residents came out to inspect the damage.

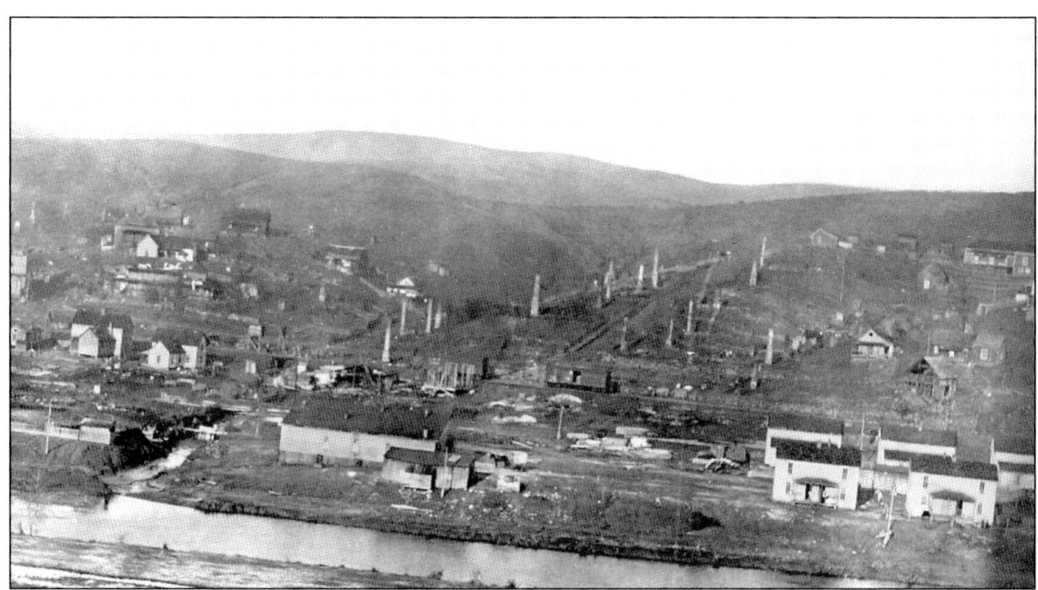

There was worse to come on the following night, when a separate fire broke out in a boardinghouse at the foot of Johnsontown Hill. The fire raged through the adjoining residential districts, leaving an estimated 400 to 500 people homeless on a cold December night. The chimneys that remain standing in this photograph mark the path of the second night's fire up Johnsontown Hill and along Water Street.

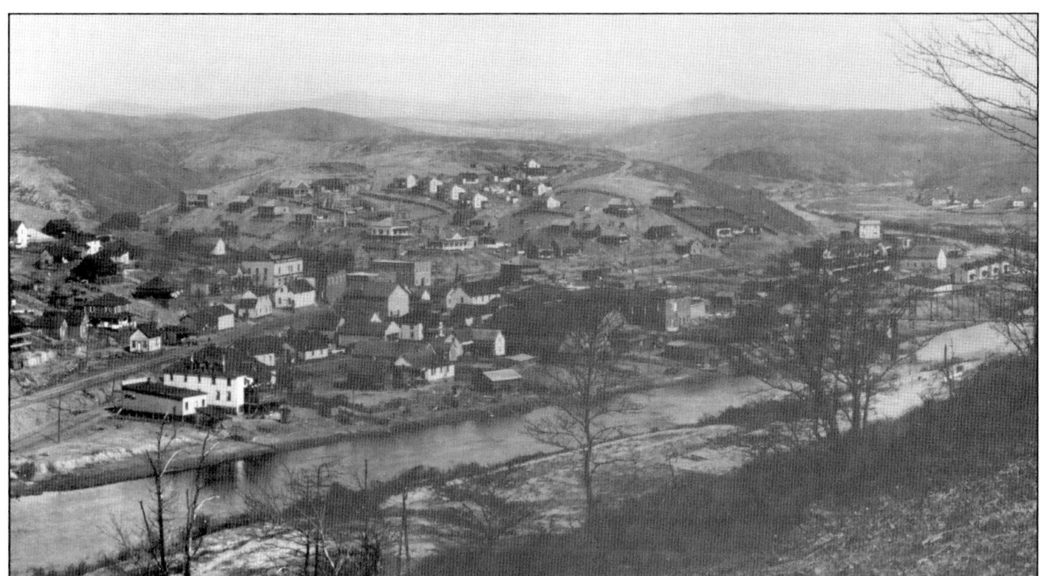

Within 10 days after the fire, Tennessee Copper Company had constructed four large new buildings near the plant to house temporarily homeless employees. Others were moved into vacant space in the foreign labor camps across the river. By the time this photograph was taken in 1911, a number of new homes on Johnsontown Hill and Water Street (top center) had already been completed, and more were under construction.

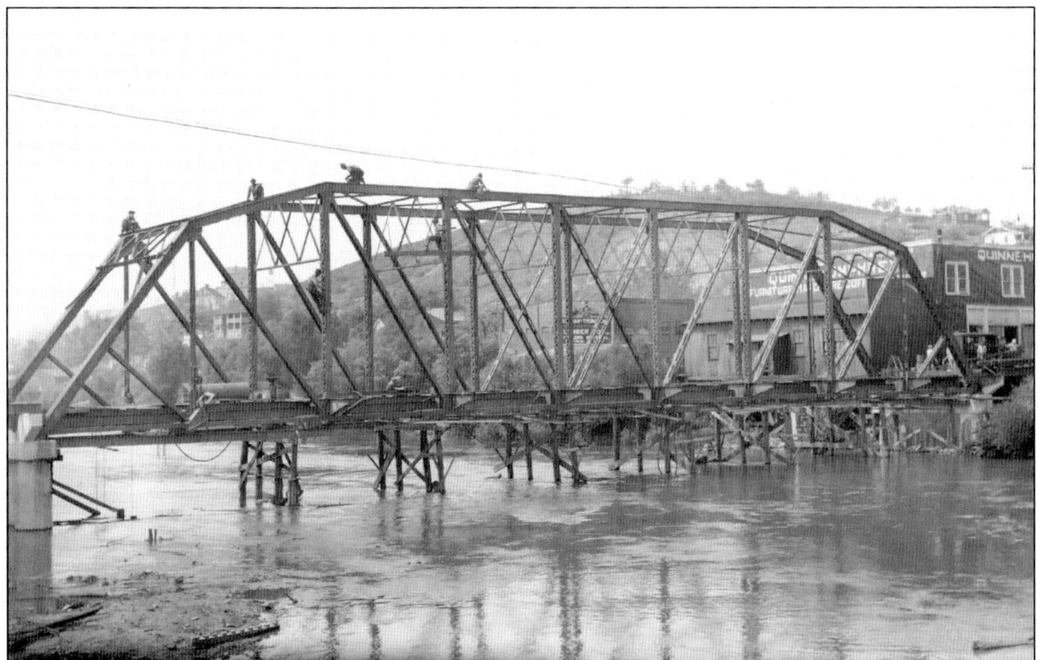

In 1911, with funding from both sides of the state line, a steel wagon bridge was built across the Ocoee River between Copperhill and McCaysville. The new bridge replaced both the footbridge and the ferry. By the 1930s, when this photograph was apparently taken, it also carried automobile traffic between the two towns.

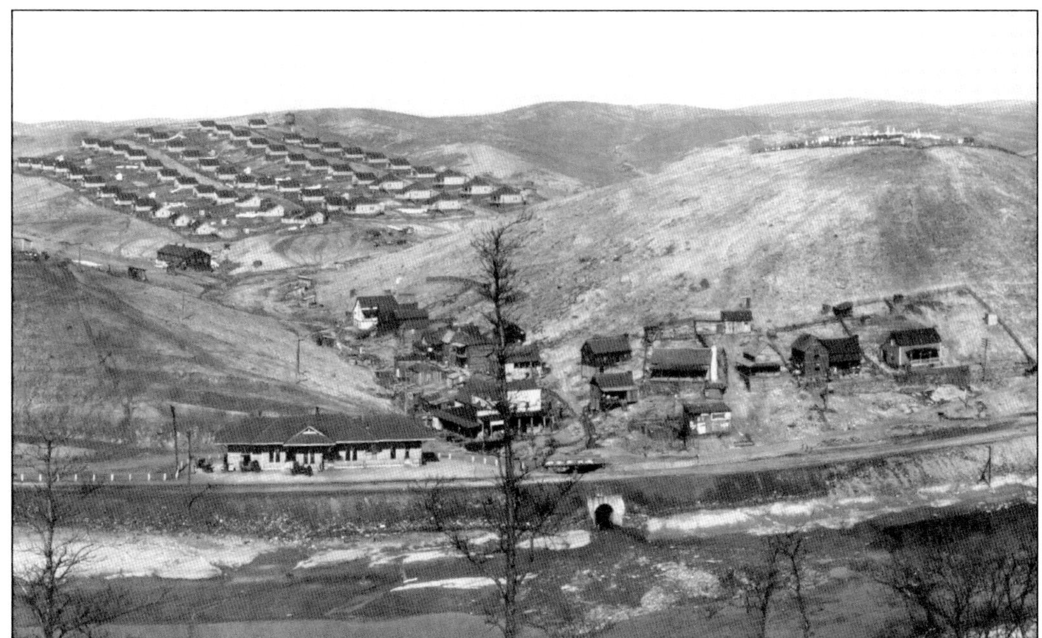

These two images, segments of a larger 1911 panoramic photograph, show the variety of Tennessee Copper Company housing at Copperhill. The workers' community of Newtown is at upper left above, and the old Methodist cemetery, more commonly known as Cemetery Hill, is at upper right. On Smelter Hill, in the image below, are the homes of TCC executives and other white-collar employees. Most houses in both neighborhoods were mail-order homes that had been precut and shipped to be erected on site. At this point, outhouses were standard even for the managerial residences on Smelter Hill. The homes overlook the TCC general office, which can be seen at far left. The two horse-drawn wagons at lower right were probably making their way to the TCC plant (not visible), just beyond the office building.

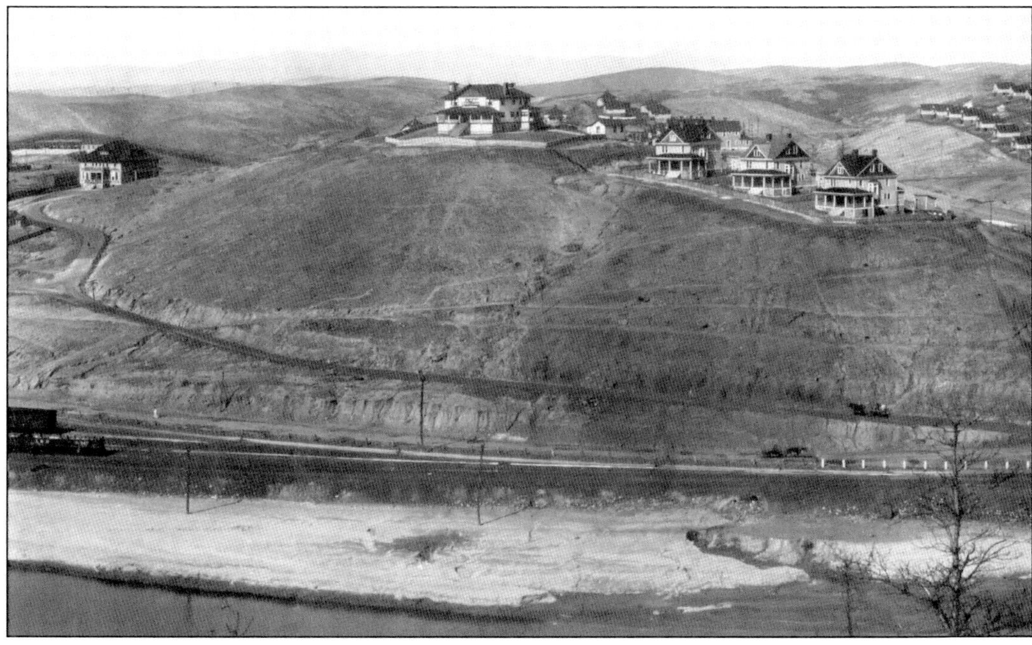

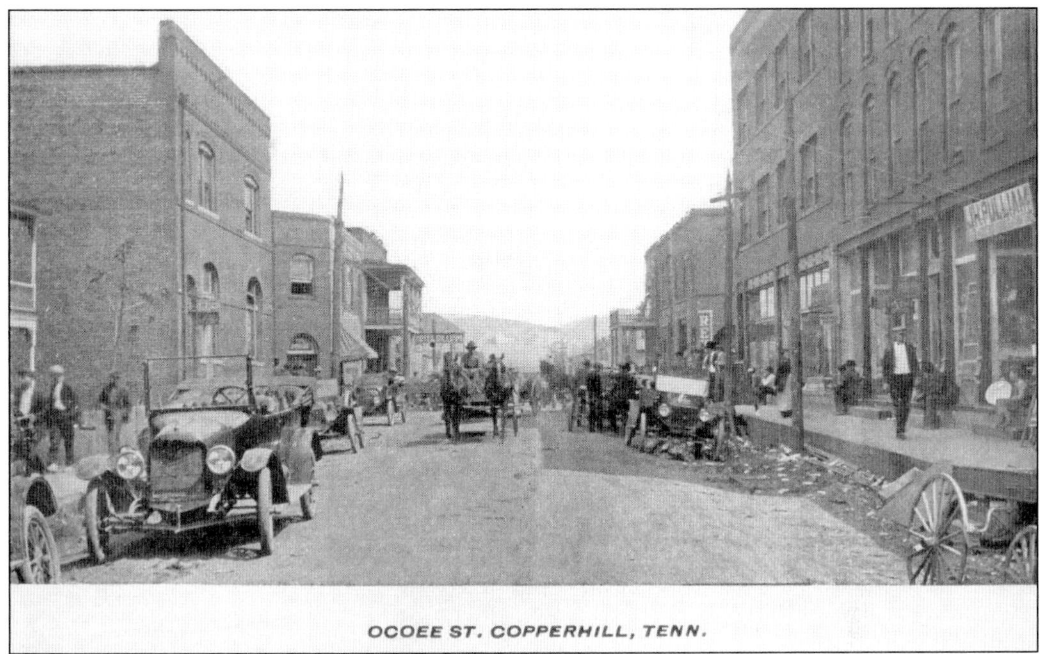

In this early 20th-century postcard view, horses and wagons still share the streets of Copperhill with the growing number of automobiles. The photograph was taken looking east along Ocoee Street, where the high sidewalks helped protect pedestrians from the dirt and mud of the unpaved street. (Author's collection.)

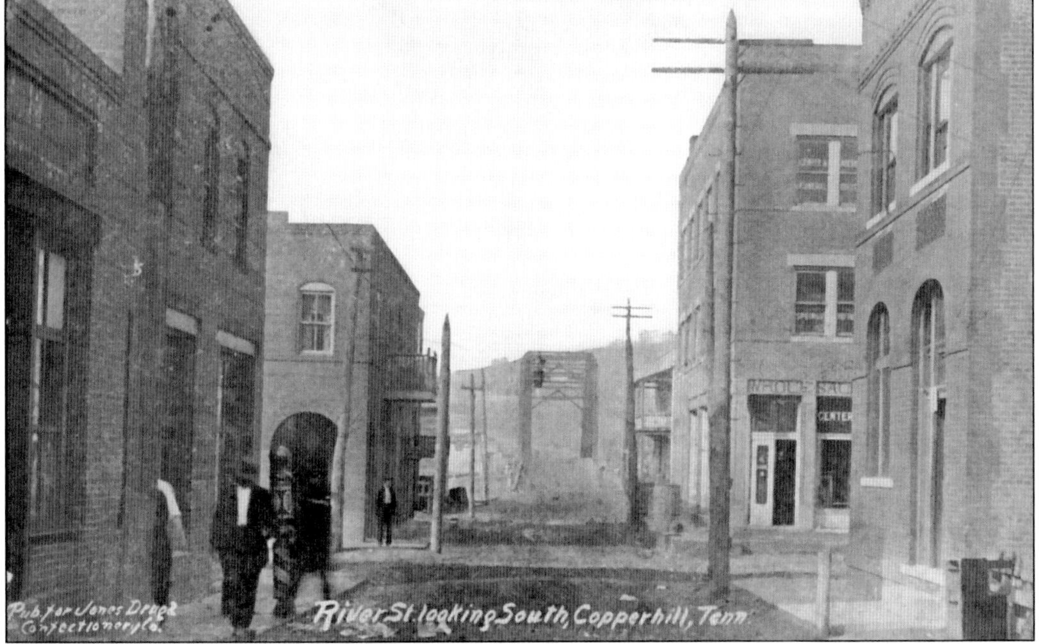

This postcard view along River Street (now Grand Avenue) in Copperhill shows the steel bridge across the Ocoee River shortly after its construction in 1911. The postcard was published for Jones Drug & Confectionery Company, which occupied the corner building on the near left, and was mailed in 1913. (Author's collection.)

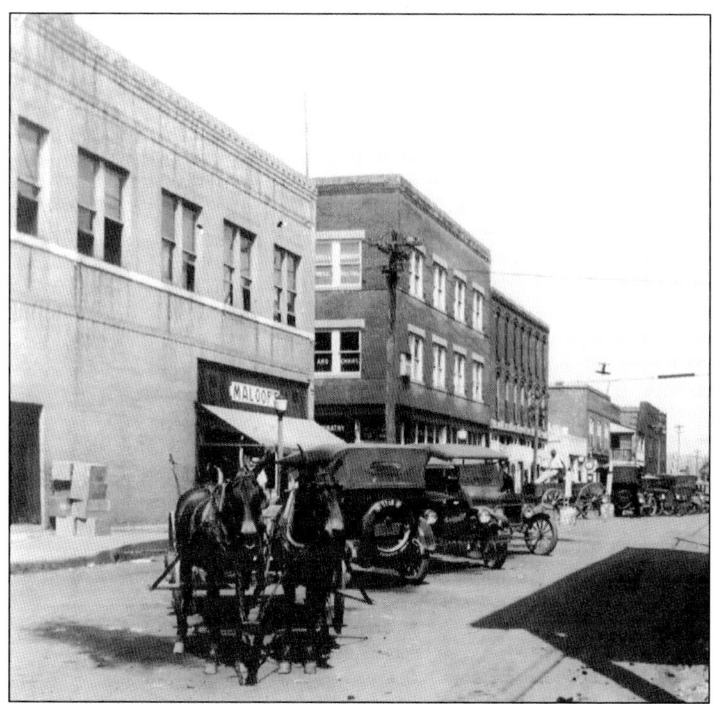

This photograph looking west on Ocoee Street was taken between 1921, when Lebanese immigrant Louis Maloof opened a department store on the southeast corner, and 1925 or 1926, when the neighboring Center & Abernathy building and several others in the same block were destroyed by fire. Maloof's was a fixture in downtown Copperhill for more than half a century, and the original building is still standing today. (Courtesy of Tri-State Electric Membership Corp.)

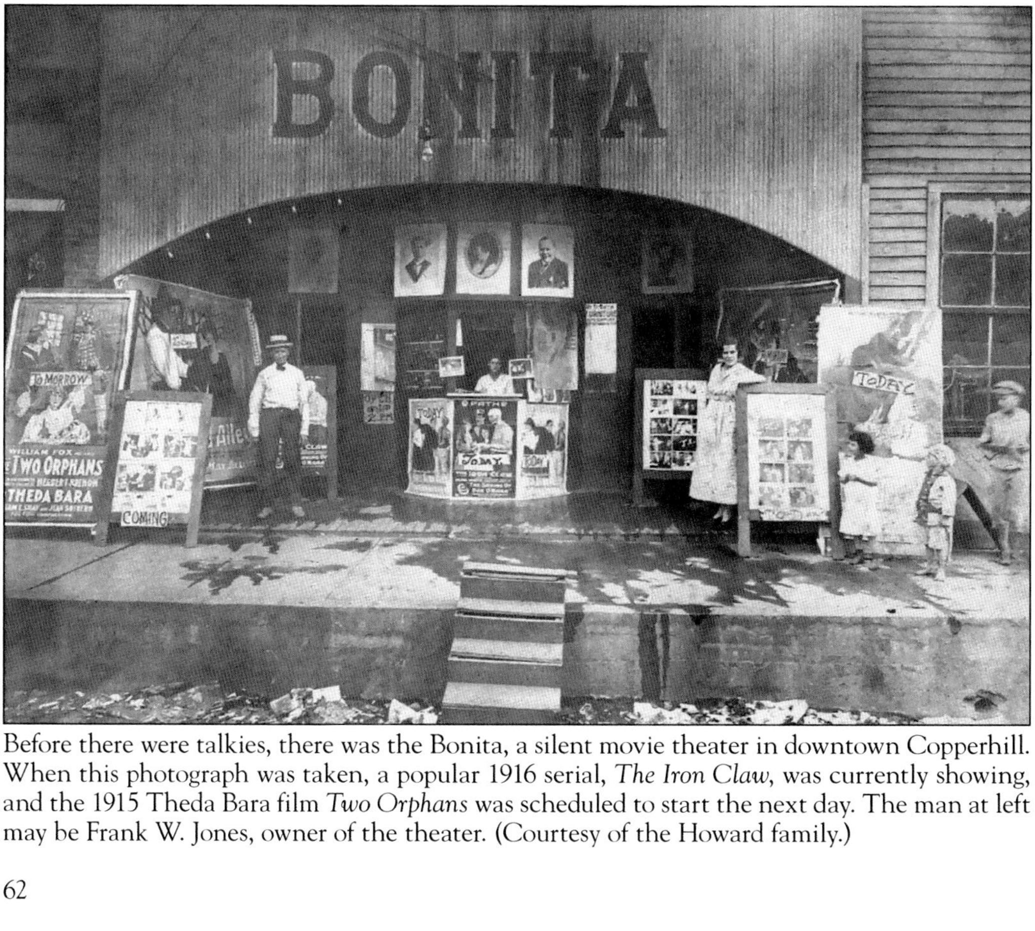

Before there were talkies, there was the Bonita, a silent movie theater in downtown Copperhill. When this photograph was taken, a popular 1916 serial, *The Iron Claw*, was currently showing, and the 1915 Theda Bara film *Two Orphans* was scheduled to start the next day. The man at left may be Frank W. Jones, owner of the theater. (Courtesy of the Howard family.)

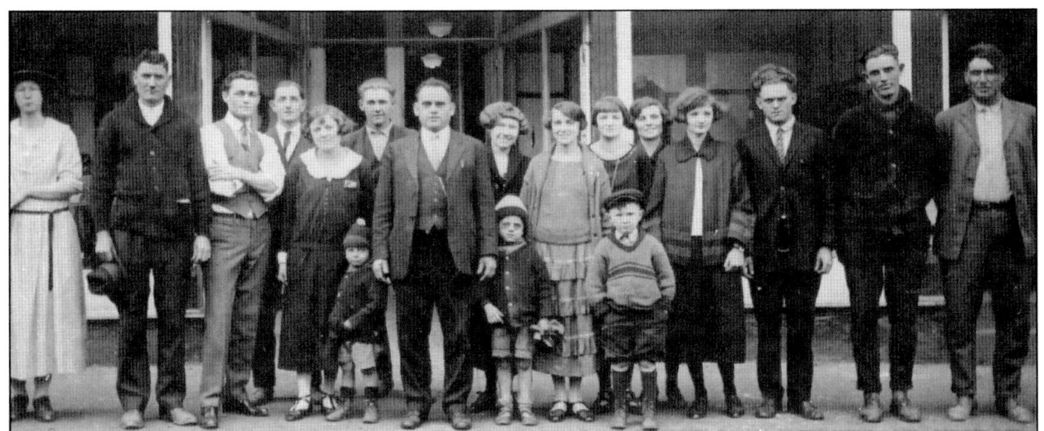

In the early 1920s, the employees of W.H. Chancey's Copperhill store gathered for the formal photograph above. W.H. Chancey, in the three-piece suit, stands near the center beside his wife, Maude, in the dark dress. The four women next to Chancey, from left to right, are Anna (Lucas) Panter, Gertie (Forrester) Frye, Thelma Casteel, and Carrie (Abernathy) Cole. Below, in a photograph taken prior to the fires of 1925 and 1926, the banner for Chancey's store can be seen next door to the Center & Abernathy building, which stood on the southwest corner of Ocoee Street. Copperhill merchant Carl Abernathy, son of Center & Abernathy co-founder George W. Abernathy, is the tall man at left in the white shirt and tie. (Both, courtesy of Greg Crawford.)

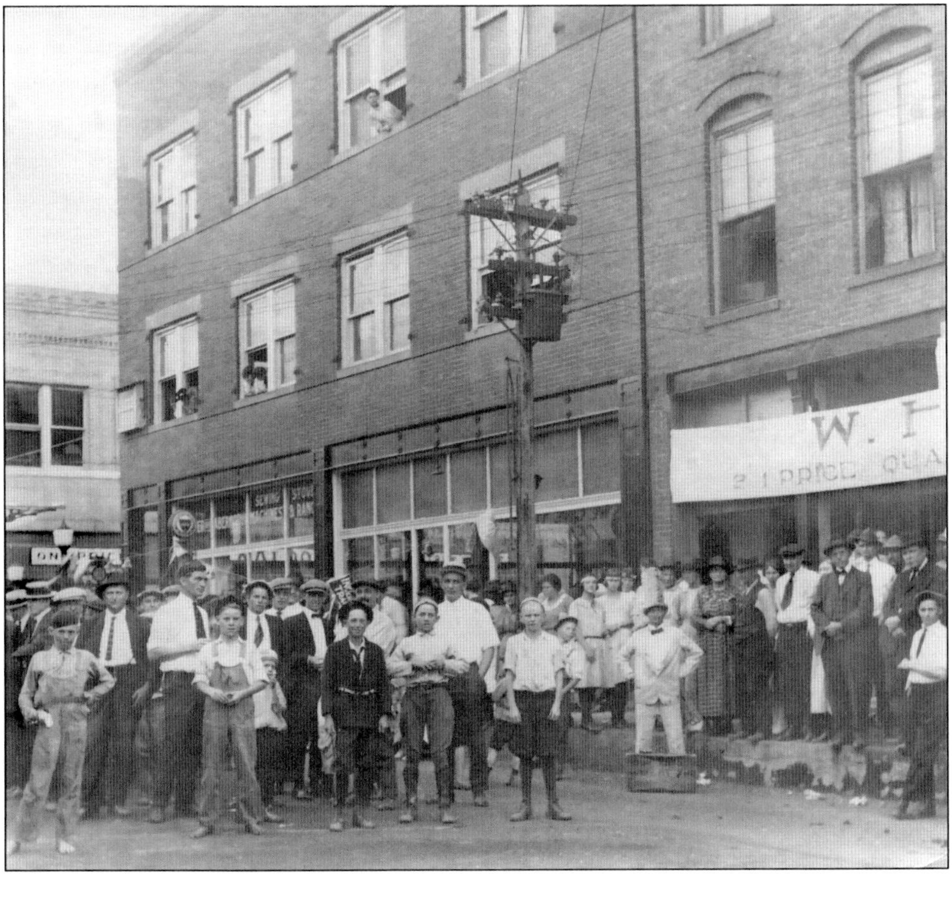

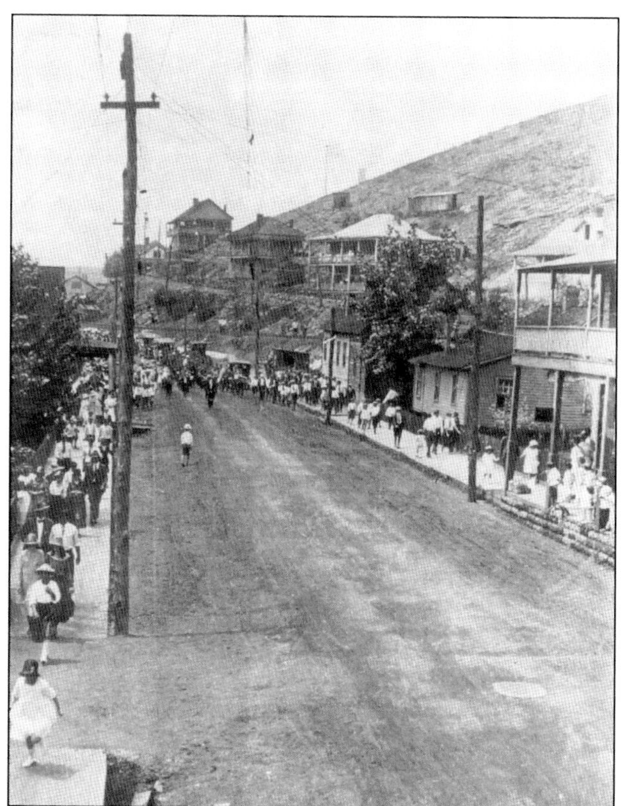

This photograph of a Fourth of July parade in Copperhill around 1923 was taken looking west along Ocoee Street toward the TCC plant, which is not visible in this view. Many of the buildings that lined the street were apparently of frame construction, and the street itself was still unpaved. Interestingly, most of the spectators on the sidewalks seem to have been walking in the same direction as the parade.

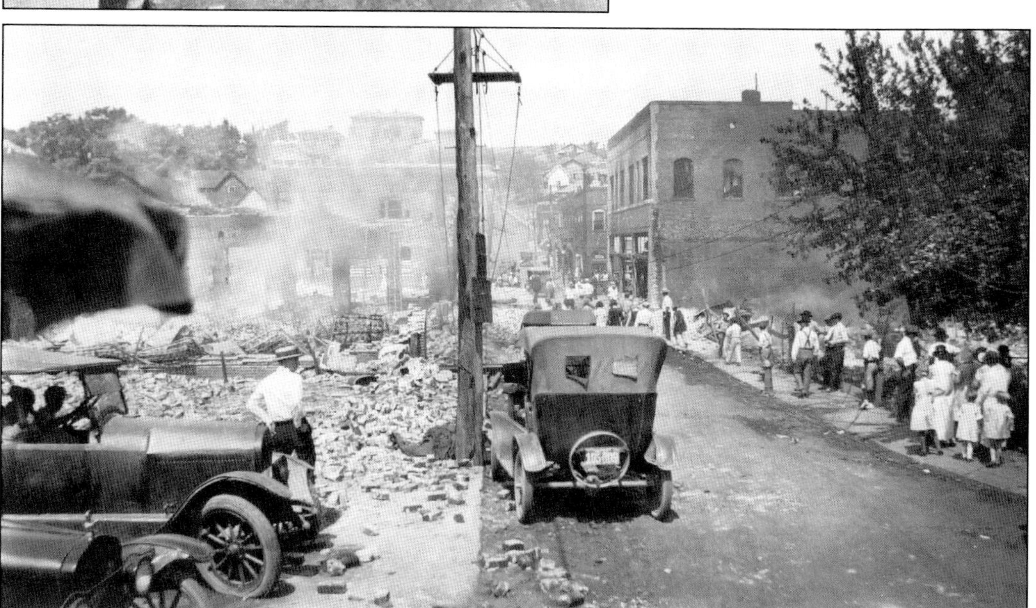

Two fires, in August 1925 and March 1926, destroyed large sections of downtown Copperhill. The Center & Abernathy building, which had stood on the left beyond the telephone pole, was reduced to rubble. Neighboring buildings fared no better. In this photograph, taken from the vantage point of the steel bridge, the Maloof building remains standing on the south side of Ocoee Street. (Courtesy of the Howard family.)

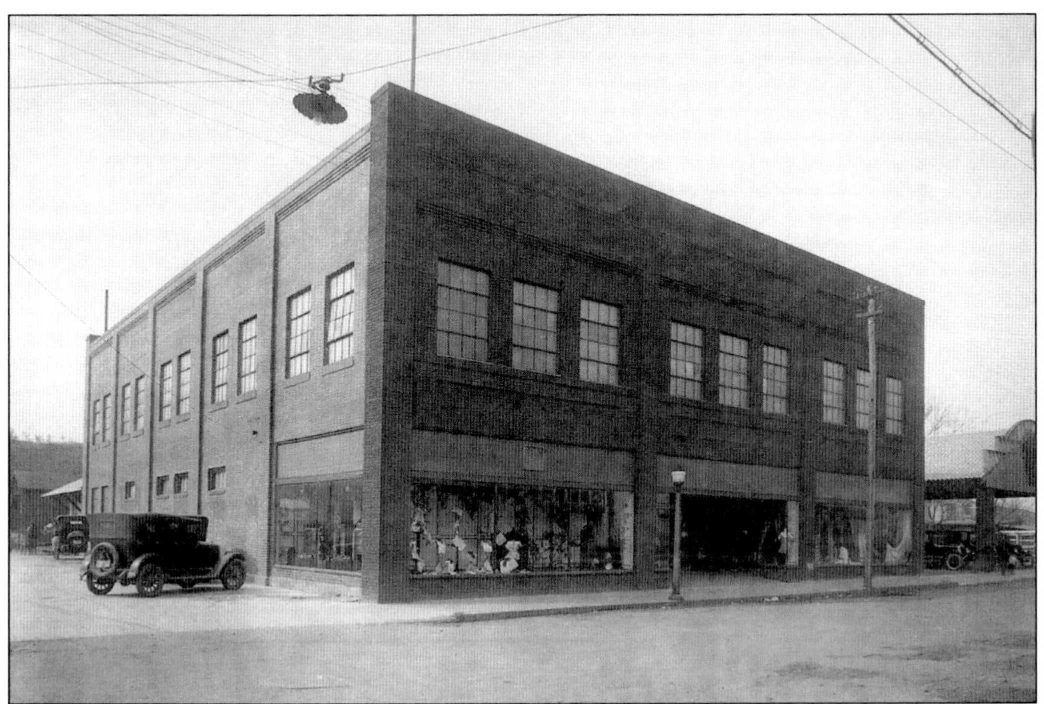

As they had done after the disastrous fire of 1910, the people of Copperhill quickly began to rebuild after the fires of 1925 and 1926. In 1926, TCC opened its newly constructed Smelter Store building (above) at the site of the present-day Copperhill Post Office. The Smelter Store stood at this location until it was destroyed by an isolated fire in 1958. The presence of the electric streetlamp in front of the new Center & Abernathy building (below), which still stands today on the site of the original building, suggests that the photograph was probably taken around 1930.

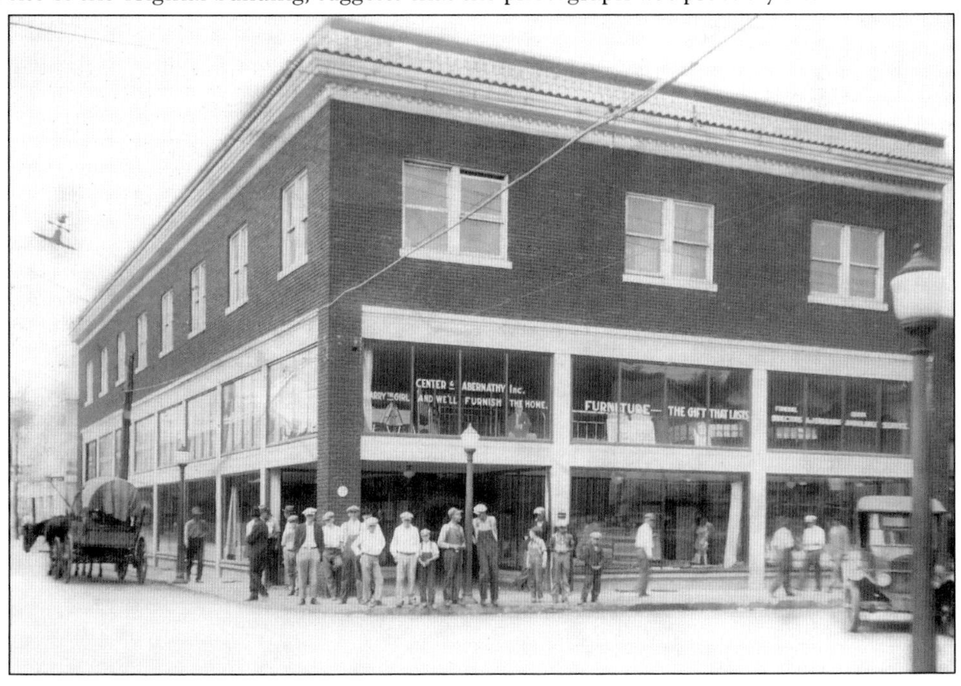

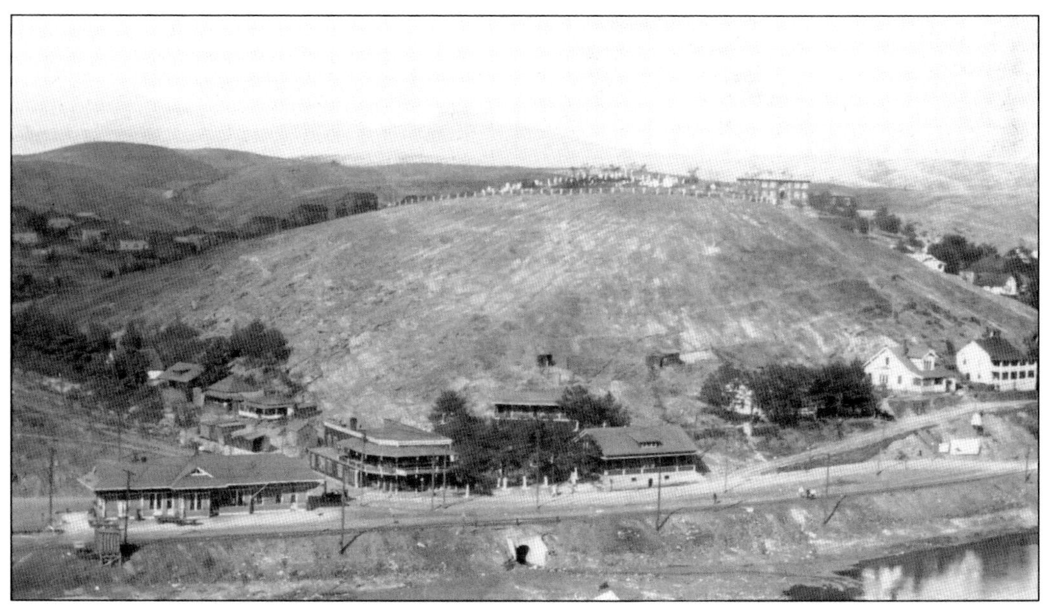

In early 1914, George Hood opened the Colonial Hotel adjacent to the L&N depot in Copperhill. "He has new furnishings throughout, hot and cold water, bathrooms, fresh spring water, steam heat, etc.," wrote the *Polk County Republican* in February 1914, "and we bespeak for him a liberal patronage." In the photograph above, which was apparently taken after the building of new TCC workers' housing on Cemetery Hill in 1916, the depot is at far left, and the hotel is immediately to its right. The brick Copperhill High School building, constructed in 1914, is on the hilltop at upper right.

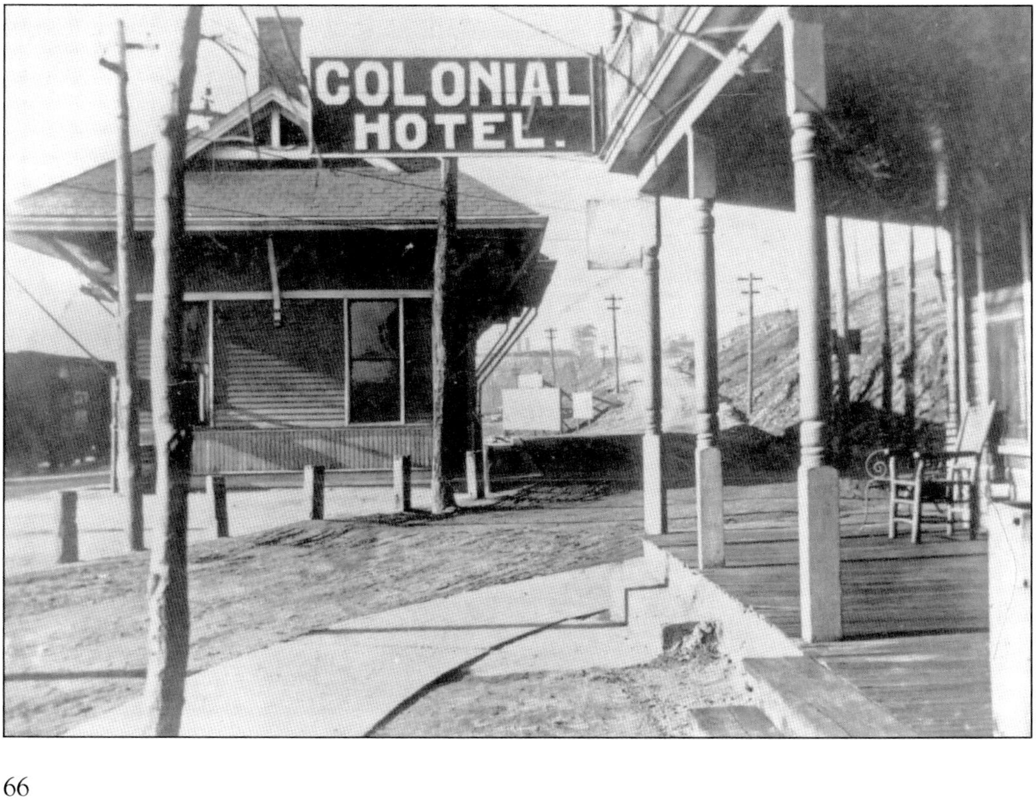

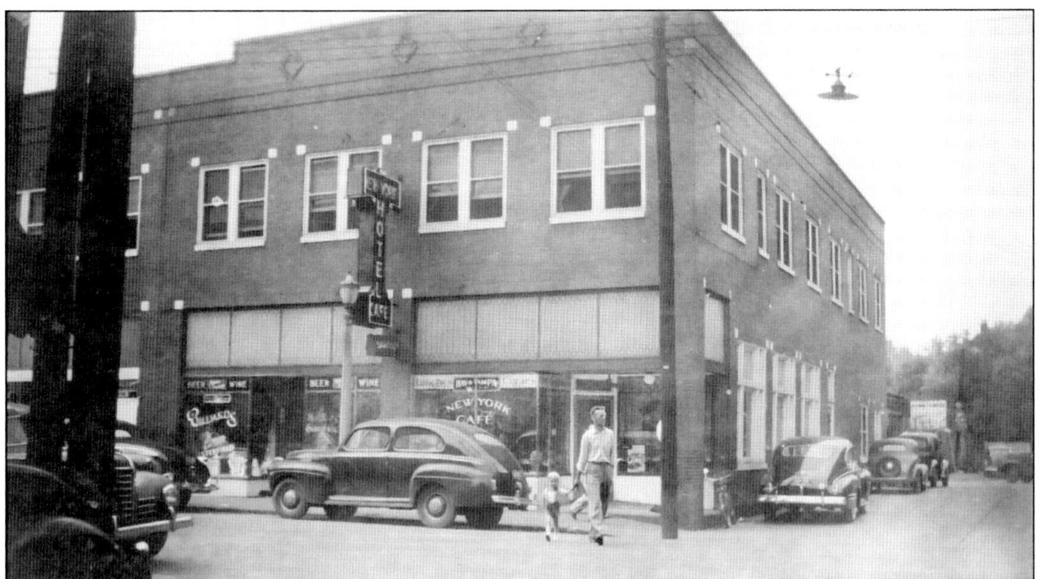

The New York Cafe in Copperhill was opened by the Gus Grevis family between 1910 and 1920. Following the fires of 1925 and 1926, it moved across the street into a new building on the corner of Ocoee and Ferry Streets. The Grevis family also ran a hotel upstairs. In the c. 1950 photograph above, Joe Hughes and his young son Steve cross the street at the intersection. Hughes was one of four men killed in 1951 in an accident at TCC's Calloway Mine. Below, Jack Bowers, whose mother Ruby worked in the cafe, stands outside the adjoining poolroom run by Gus Grevis's oldest son, George. The Grevises were Copperhill's first permanent Greek residents.

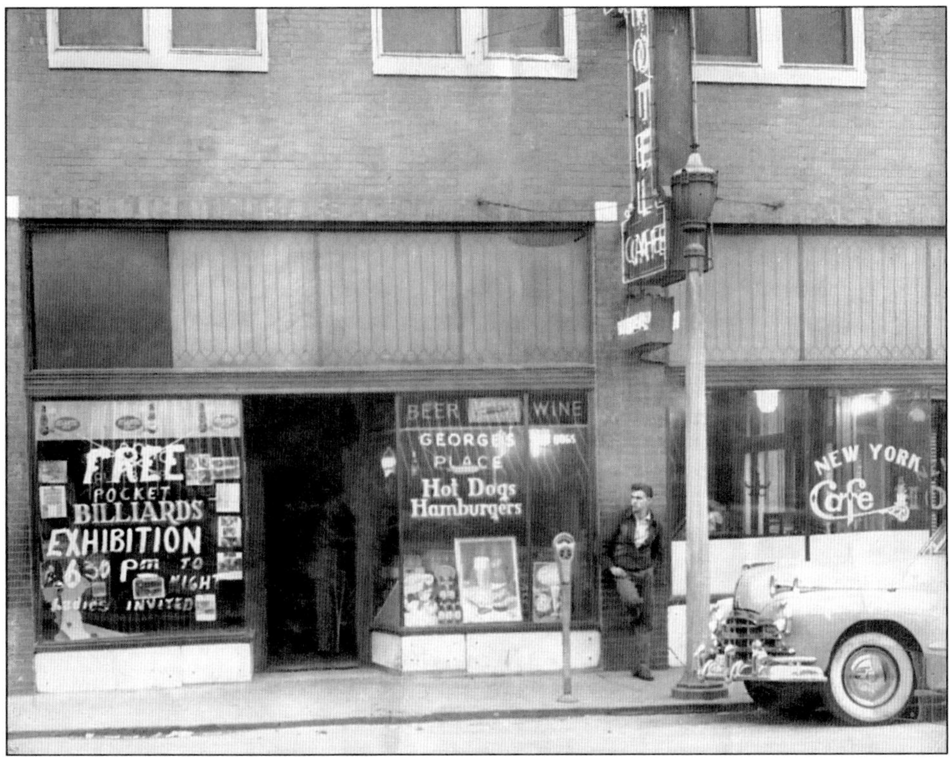

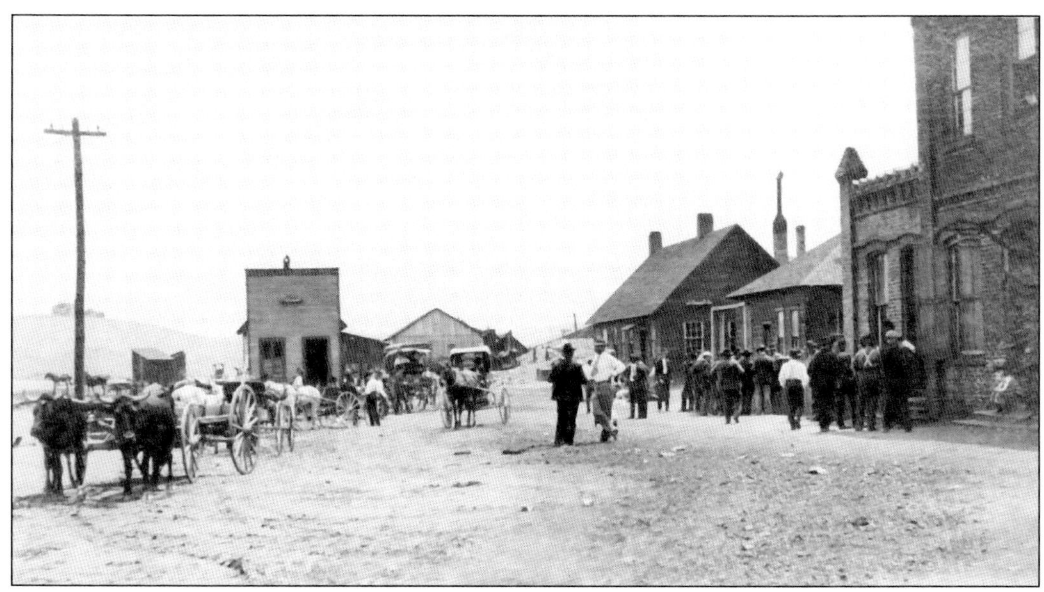

In these two early 1900s photographs, apparently taken at the same time but from different angles, the Isabella Post Office is on the left, and the company store built by J.E. Raht in 1870 is on the far right. Above, A.S. Mull's livery stable stands to the right of the post office, and an oxcart waits for its driver in the left foreground. More of the company store's impressive brick facade is visible below, and Raht's old Union Church can be seen on a nearby hilltop. DSC&I's Isabella plant was out of frame just beyond the store building. From this main business district and the adjoining plant facilities, Isabella branched out into what were essentially two separate communities, with company housing strung along one hollow and the homes and stores of Stuarttown along another.

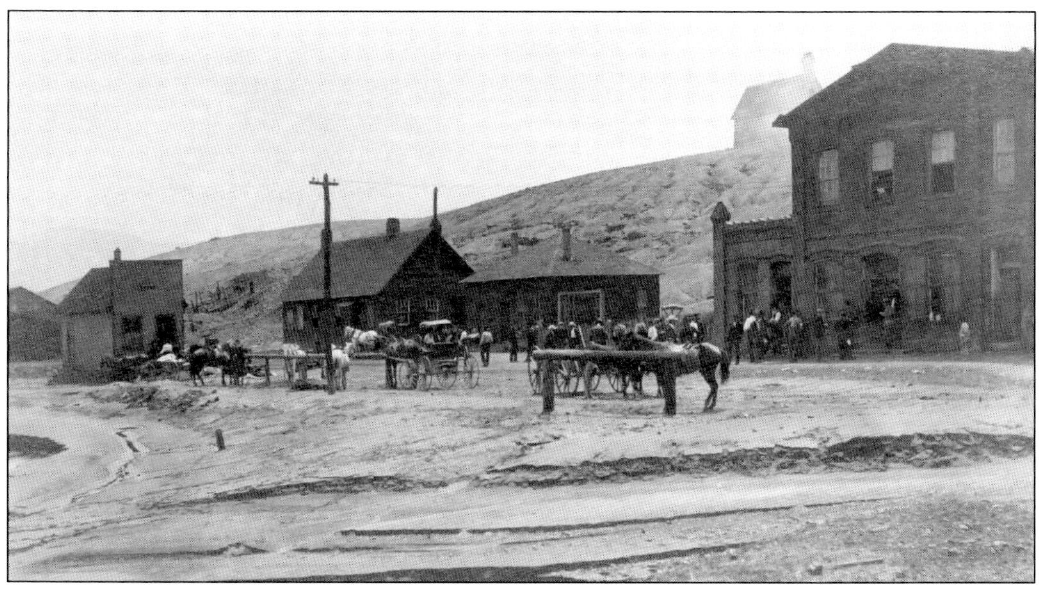

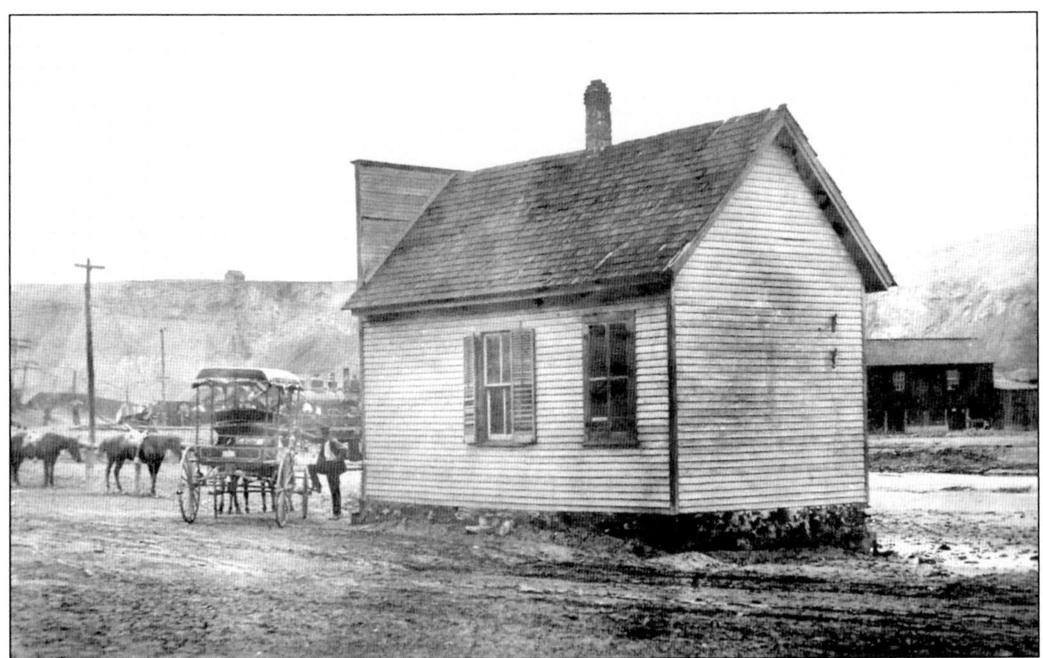

The post office at Isabella was established in 1893, four years after the purchase of the Isabella mining properties by DSC&I. In this c. 1900 photograph by Thorn Smith, the horse-drawn mail wagon waits in front of the building.

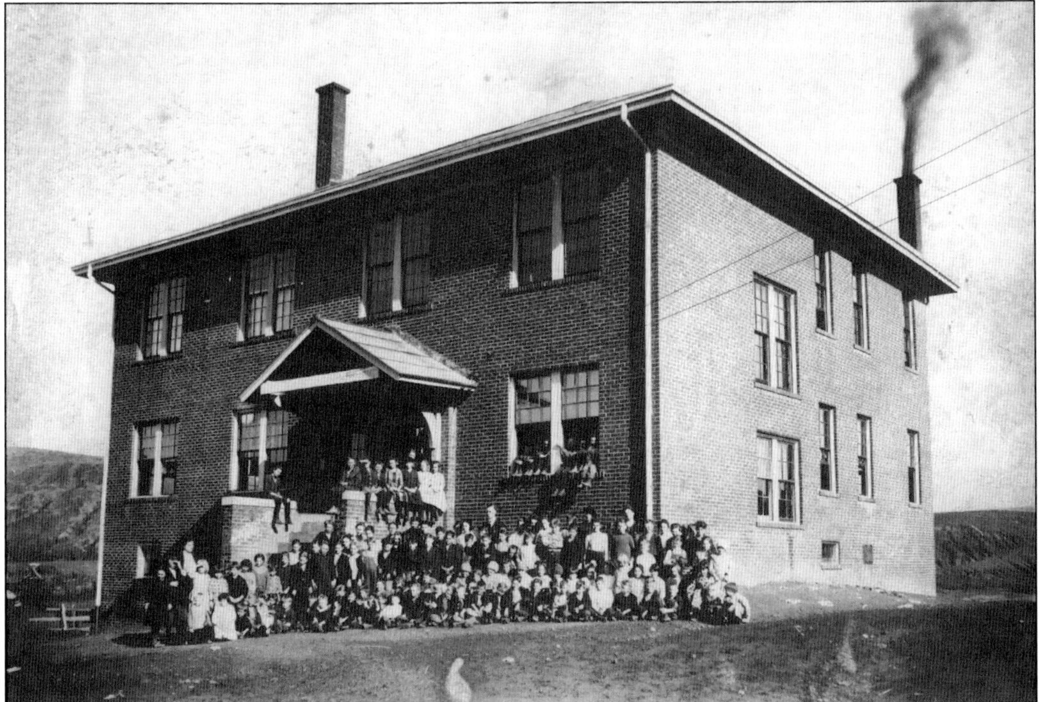

On May 25, 1917, DSC&I assistant general manager W.F. Lamoreaux ran an ad in the *Polk County Republican* soliciting bids for the erection of a new grammar school building at Isabella. In this undated photograph, students and teachers pose outside the completed building.

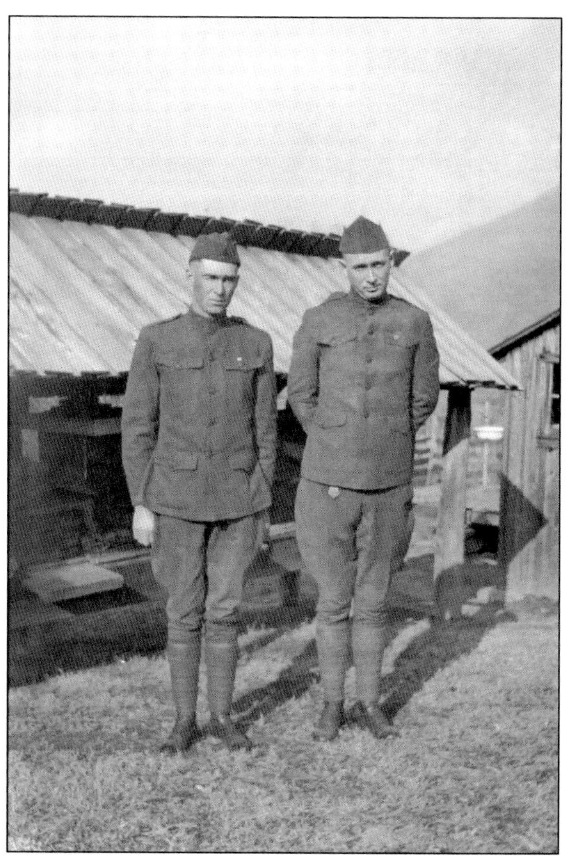

Lucius (left) and Ananias Burger, two of the five sons of Alfred and Candace Burger, pose in front of their family's log cabin at Isabella shortly after their return from World War I in 1918. The two young soldiers had both been born in the Burger cabin, which was built in 1804 by their grandfather Adam Burger and is still standing today. (Courtesy of Thomas Panter III.)

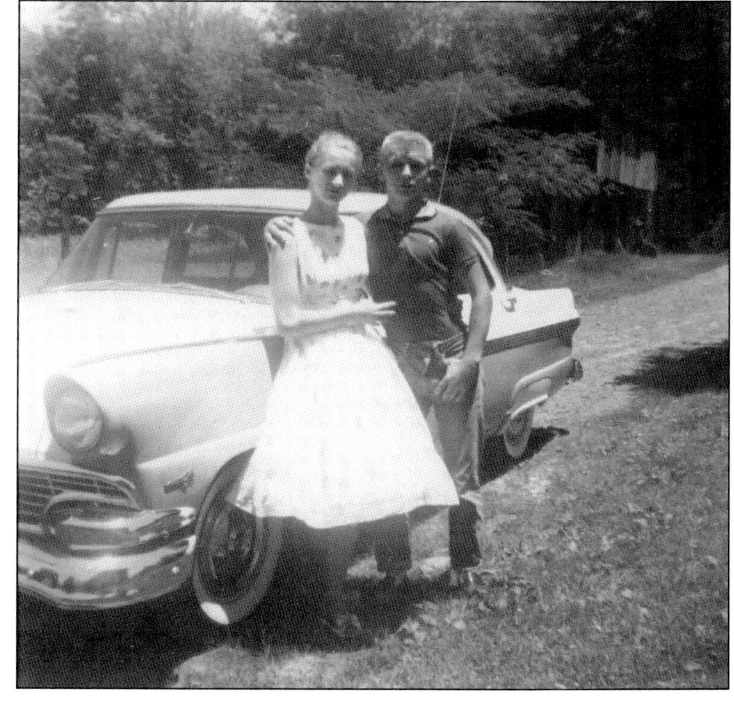

Nearly 40 years later, Lucius Burger's daughter Lynette and her future husband, Thomas Panter Jr., posed for this snapshot outside the Burger cabin at Isabella. The 1956 Ford behind them was just one of many prizes that had been won on *The Price is Right* by Thomas Panter Sr., father of Thomas Jr., who still holds the record for the number of wins (11) on the television game show. In 1959, the tan and white Ford would carry Thomas and Lynette Panter on their honeymoon. (Courtesy of Thomas Panter III.)

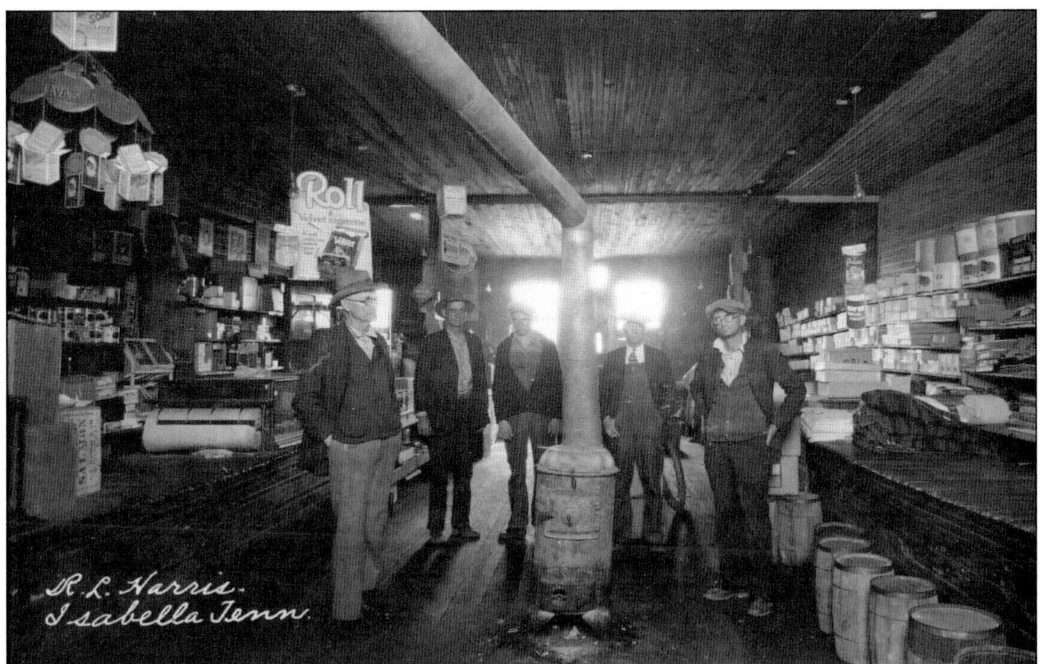

In the early part of the 20th century, Isabella was sufficiently prosperous to support several merchants. Identified in the c. 1915 photograph above are Robert L. Harris, his son Clyde Harris, and Lee Magbie. Clyde Harris, who would have been about 18 years old, is probably the younger man at center. Merchandising materials for Lava soap, Carnation milk, Rumford baking powder, and Velvet cigarettes hang from the ceiling. Both Lucius Burger and his brother Ananias were enumerated in the 1930 census of Isabella as dry goods salesmen; Ananias Burger may be the older man in the photograph of A. Burger & Brother below, but the younger man is unidentified.

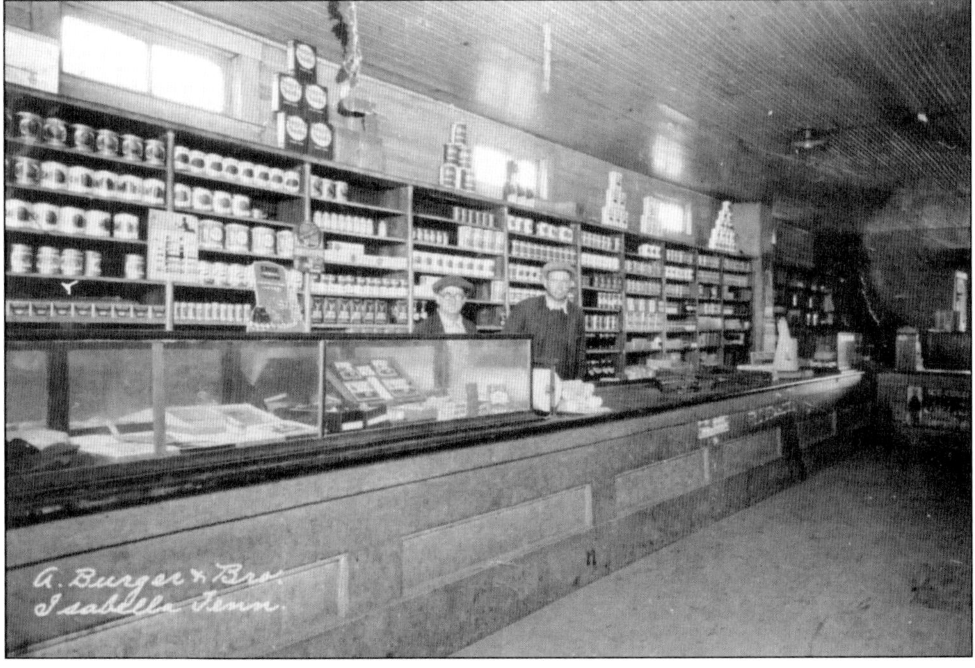

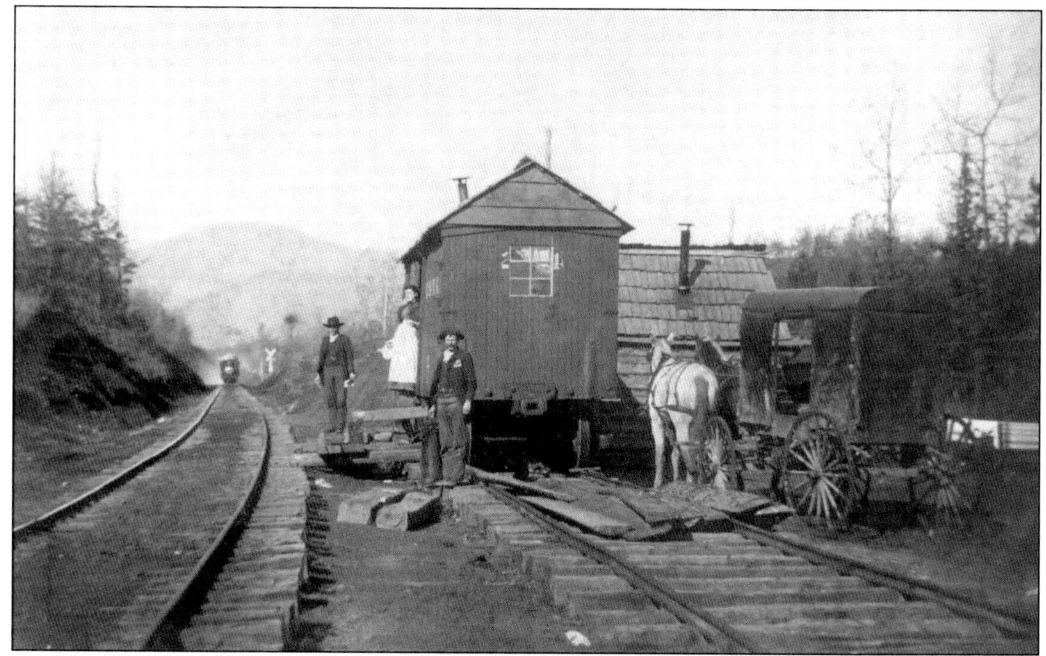

With the coming of the railroad to the Ducktown Basin, the village of Simtown, about a mile west of present-day Ducktown, was chosen as the most accessible location for the new Ducktown station. In this 1891 photograph, mail carrier John Hyatt (on the steps) waits for the arriving train. In 1899, a post office was opened at the station, and the village was renamed Postelle in honor of Ducktown physician J.M. Postelle.

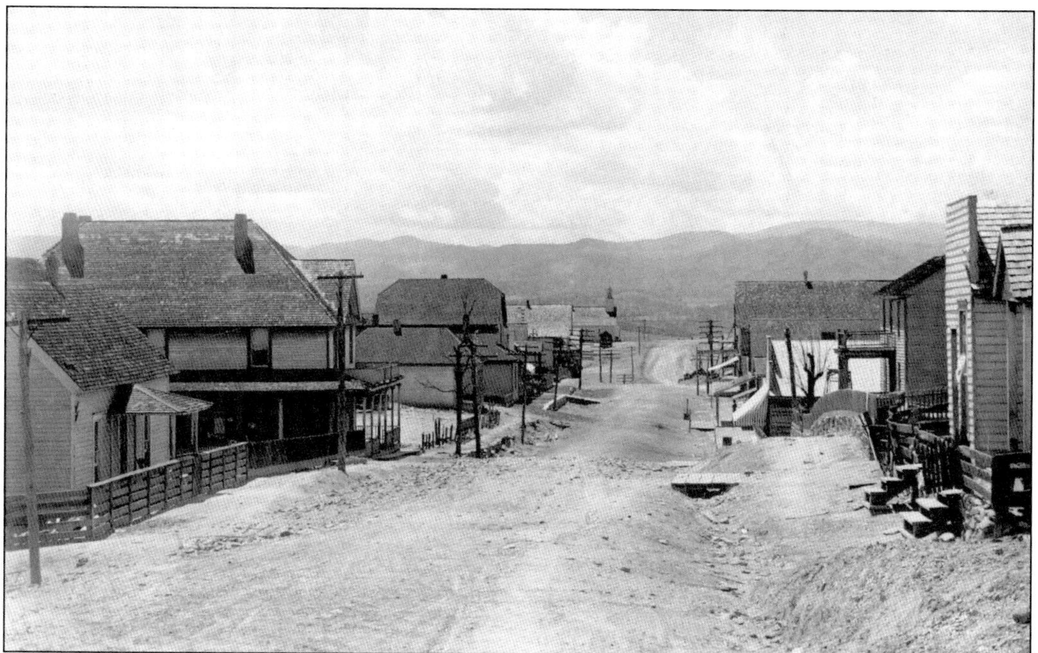

E.H. Westlake took this photograph of Ducktown's Main Street in 1904. With the return of the copper mines, the formerly sleepy little town had begun to grow and prosper again, although sidewalks and paved streets were still well into the future.

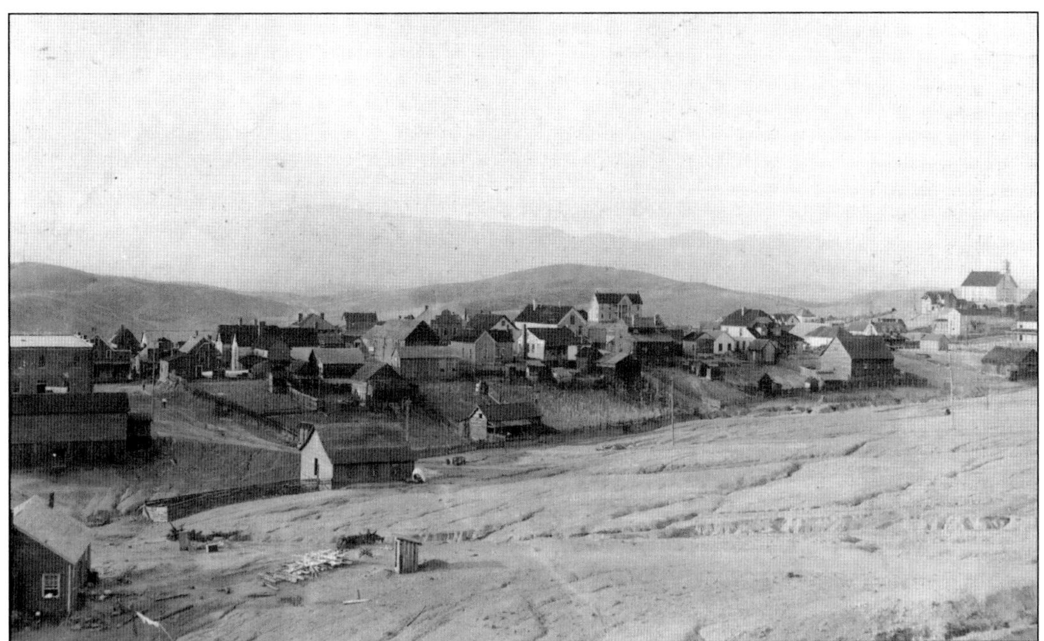

In this view of Ducktown, photographed by E.H. Westlake in 1905, the original Methodist church can be seen on the hilltop at far right. Locals had dubbed the location "Happy Top," because it was said that the joyful singing of hymns at the church could be heard throughout the town. The church was originally the Methodist Episcopal Church, South; for a while, Ducktown also had a Methodist Episcopal Church, North.

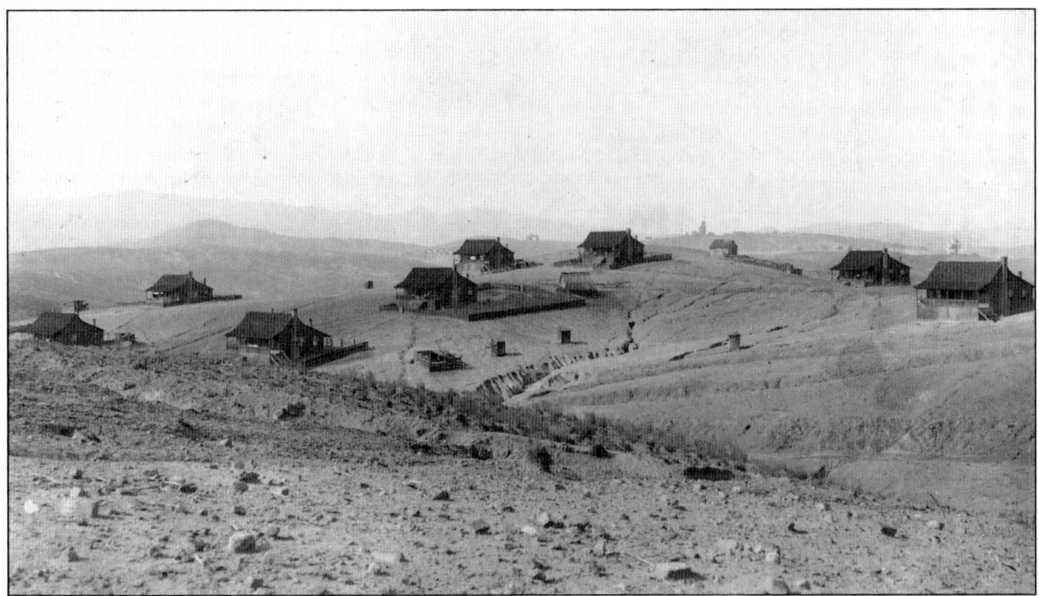

Workers' cottages at London Mine straggle across the bare hills northeast of Ducktown in this 1905 photograph by E.H. Westlake. The Burra Burra headframe can be seen faintly in the distance at center right. The cottages appear to have had identical floor plans and may have been mail-order houses, typical of worker housing in this era.

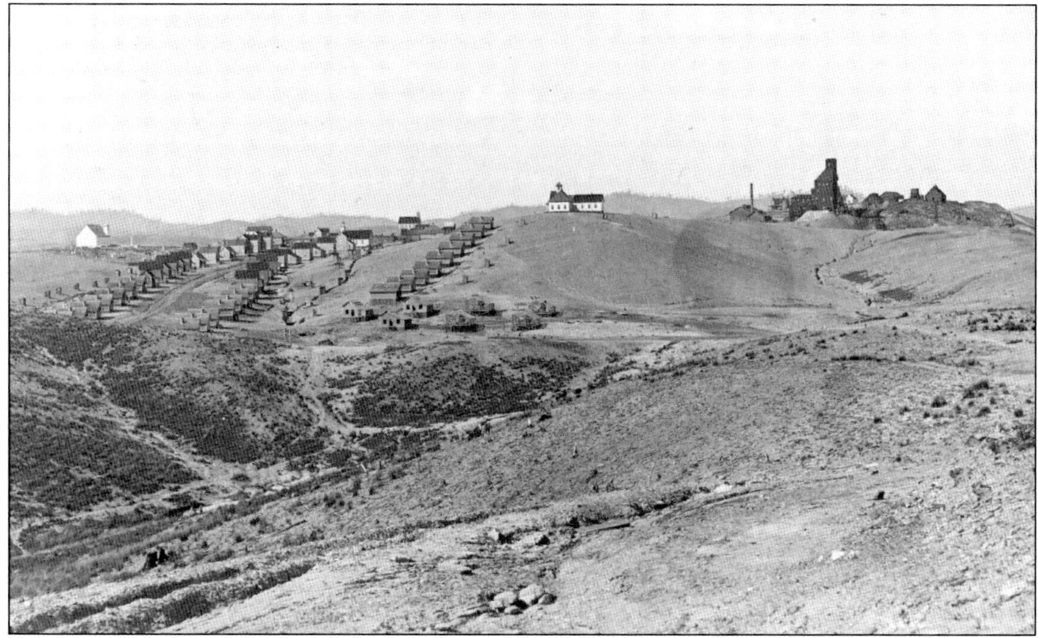

This 1906 E.H. Westlake photograph of Ducktown shows neat rows of workers' housing, with the Methodist Episcopal Church, South and cemetery at far left and the Burra Burra mine complex on the hilltop at right. The building on the hill at the center appears to be the early frame schoolhouse that was replaced a few years later by a modern brick high school building. The other church that can be seen in the distance beyond the houses may be the Methodist Episcopal Church, North.

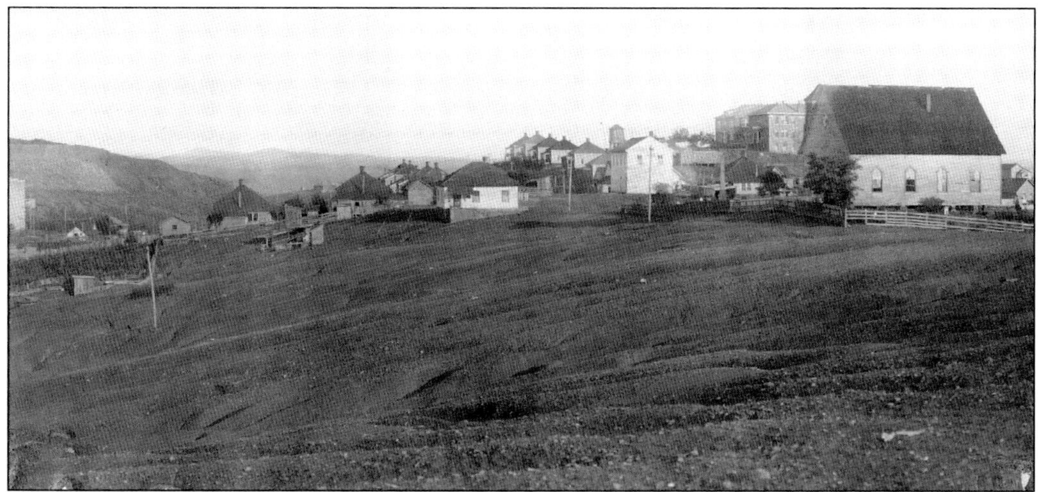

This undated photograph of Ducktown was taken sometime after the completion of the brick grammar school in January 1918. Both the grammar school and the high school, which had been constructed a few years earlier, stand side by side on the hill beyond the Methodist church at right. The tall obelisk of the Kilpatrick monument can be seen to the left of the church building, which appears to have lost its steeple at this point, but most of the cemetery is not visible due to the camera angle.

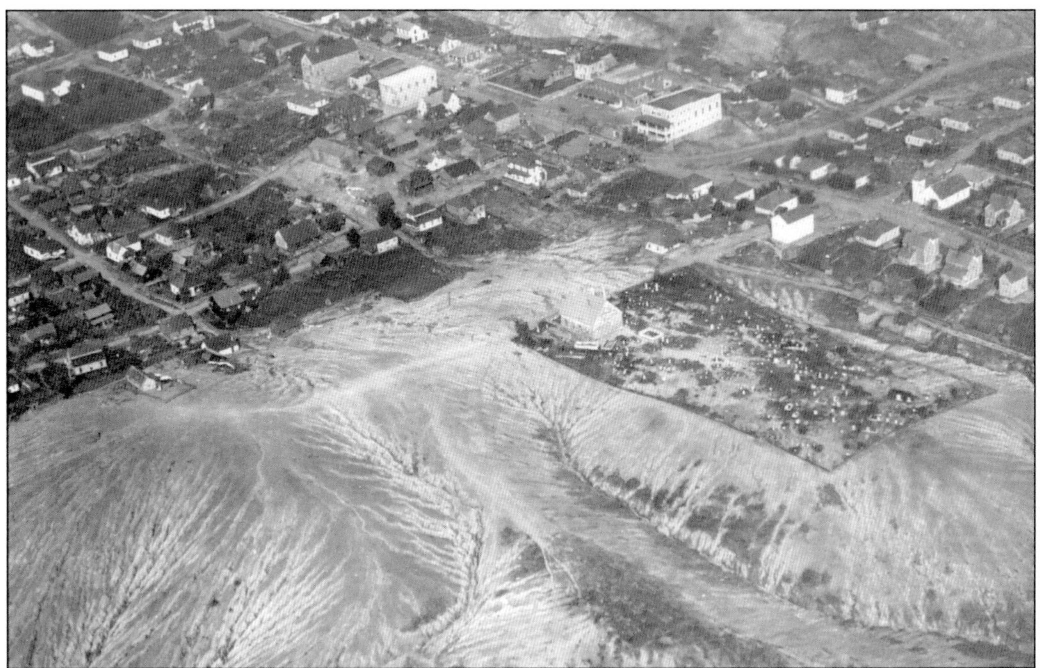

These aerial photographs document some of the changes in the city of Ducktown in the first half of the 20th century. Above, the Methodist church sits in its original location at the edge of the cemetery. The YMCA building, which was constructed in 1911 and enlarged by the addition of a third floor in the 1920s, is the three-story white building at upper right center. The old Mine City Baptist Church is the white building with a tower at the right edge. In the postcard image below, the Methodist church has been moved across the street to its current location, and the once-barren landscape is punctuated with trees that partially obscure the YMCA building. The current Mine City Baptist Church can be seen at far right.

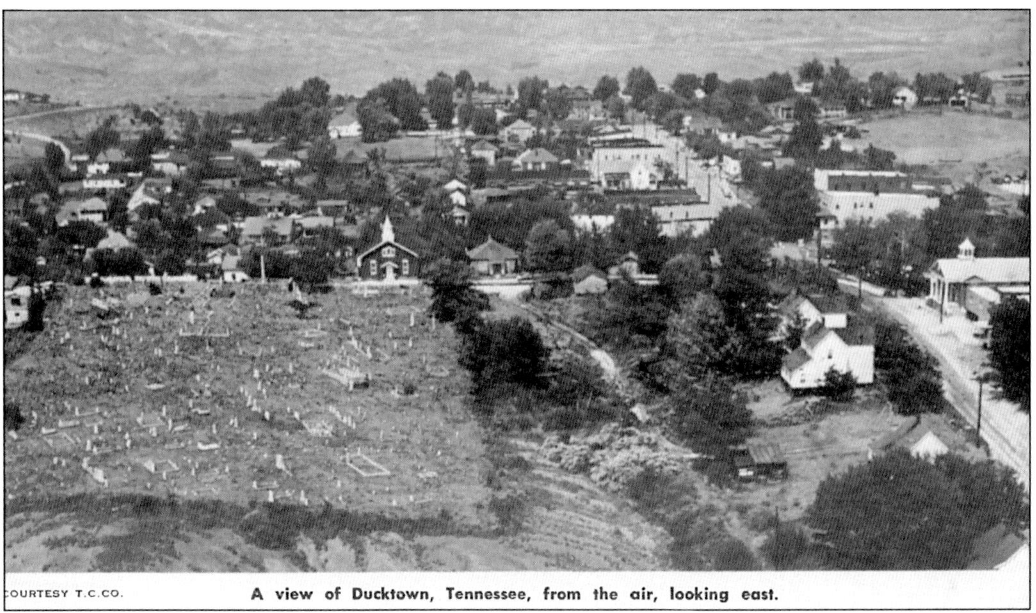

COURTESY T.C.CO. **A view of Ducktown, Tennessee, from the air, looking east.**

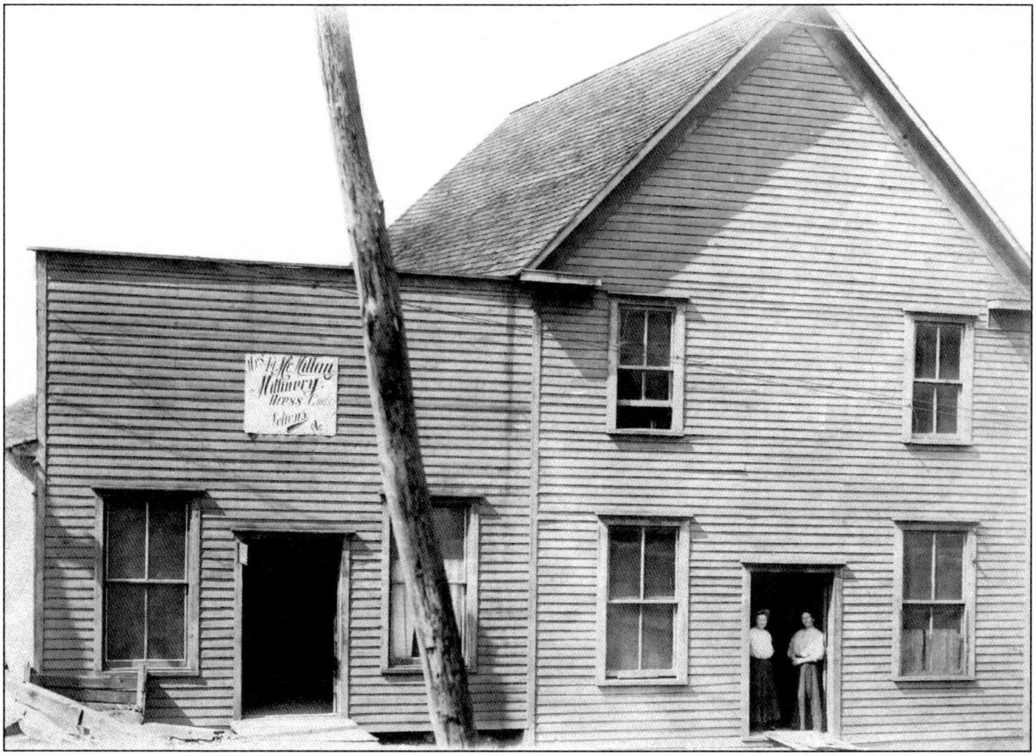

Mrs. F.L. McMillan's Millinery Company, above, advertised "Dress Goods, Lotions, etc." Its owner was apparently Salina McMillan, who was enumerated as a milliner in the 1910 census of Copperhill. Her husband, Frank McMillian, was listed as a railroad engineer. Below, clothing and luggage are featured in the windows of Millen Brothers Bargain Store; a horse-drawn cart waits in front of the store, and a pig roots placidly in the side yard. Both stores are thought to have been located in Ducktown. Judging from clothing and surroundings, the photographs appear to date from the very early 1900s. (Both, courtesy of Greg Crawford.)

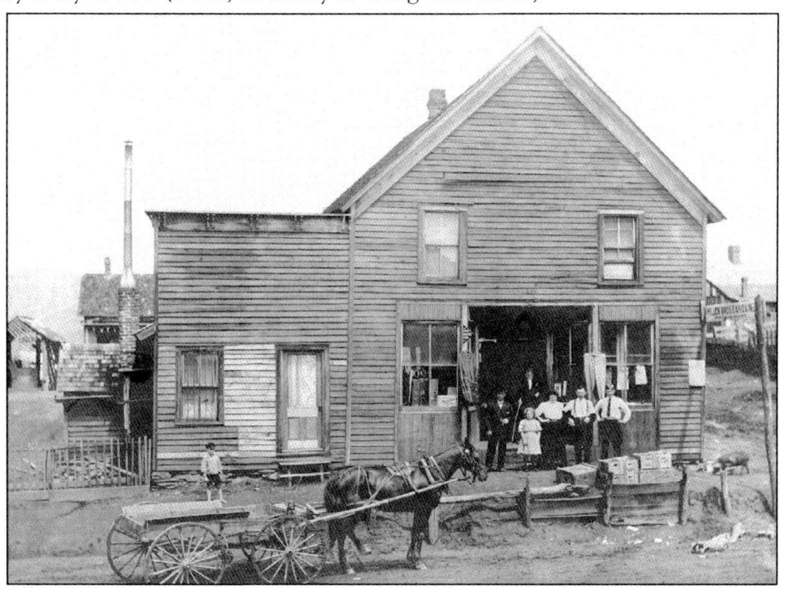

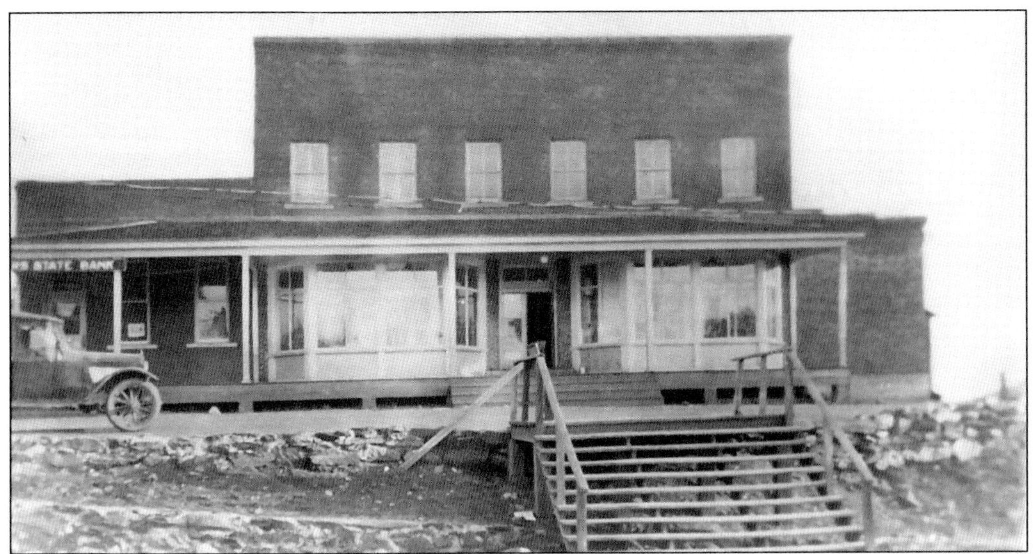

The First National Bank of Ducktown, chartered in 1909, was succeeded in 1917 by the Miners State Bank, which can be seen at left in this photograph of the first Tennessee Copper Company store at Ducktown. This photograph dates from the period between the chartering of Miners State Bank in 1917 and its liquidation in 1923.

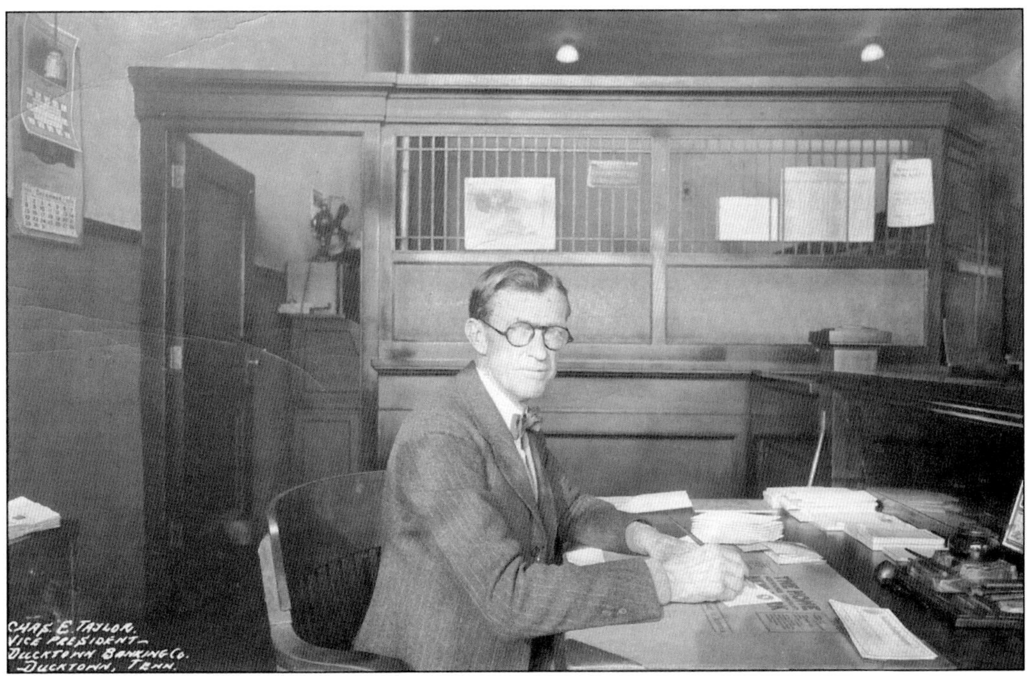

This photograph of Charles Taylor of the Ducktown Banking Company probably dates from the 1930s. Taylor joined the bank in 1904, was serving as cashier by 1914, and had been promoted to vice president by the time this photograph was taken. In 1942, he succeeded the late Dr. L.E. Kimsey as president of the bank.

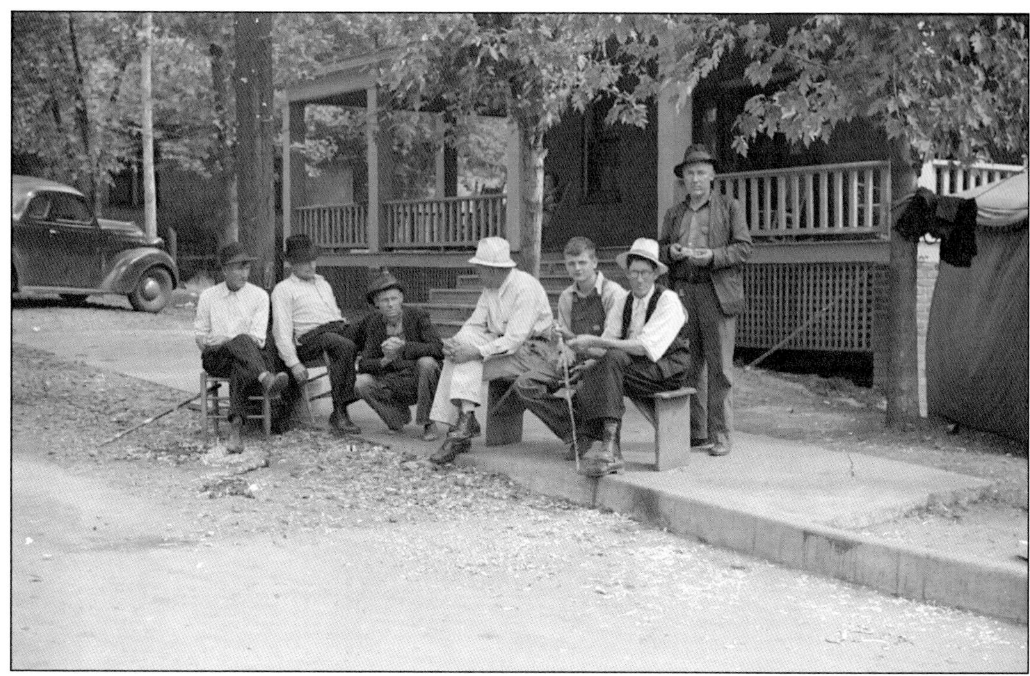

The Ducktown Hotel, opened in the early 1900s, was in the downtown business district within sight of Burra Burra Mine. In these two 1939 photographs by Farm Security Administration photographer Marion Post Wolcott, union strikers outside the hotel wait for strikebreakers leaving the mine. Identified in the photograph below, from left to right, are Bob Rymer, Everett "Flathead" Goode, Wesley Davis, and Burt Hensley, who are also among the strikers in the larger group above. Below, the Burra Burra shop building and ore bins are visible above the hotel sign. (Both, courtesy of the Library of Congress.)

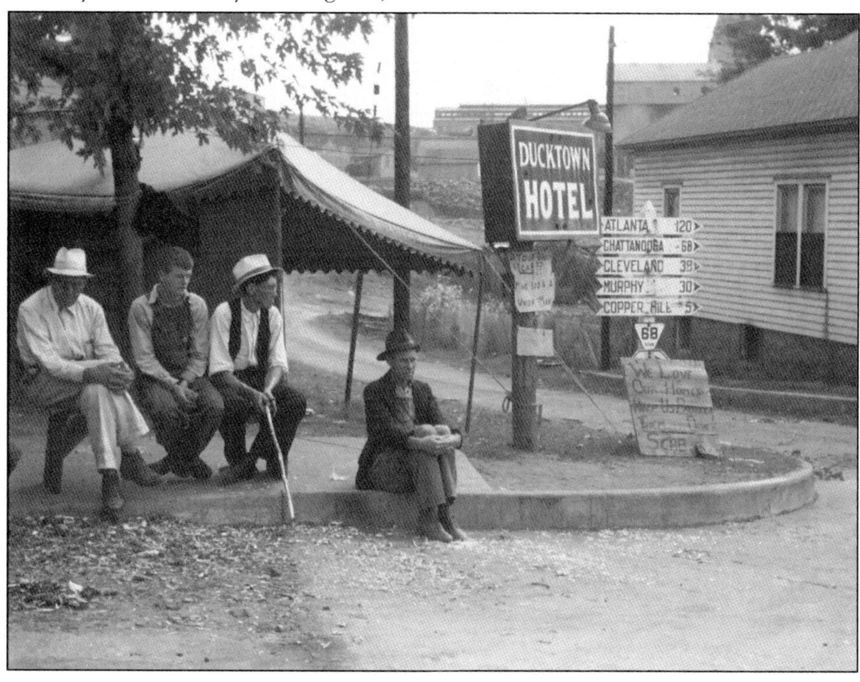

Three

People and Pastimes

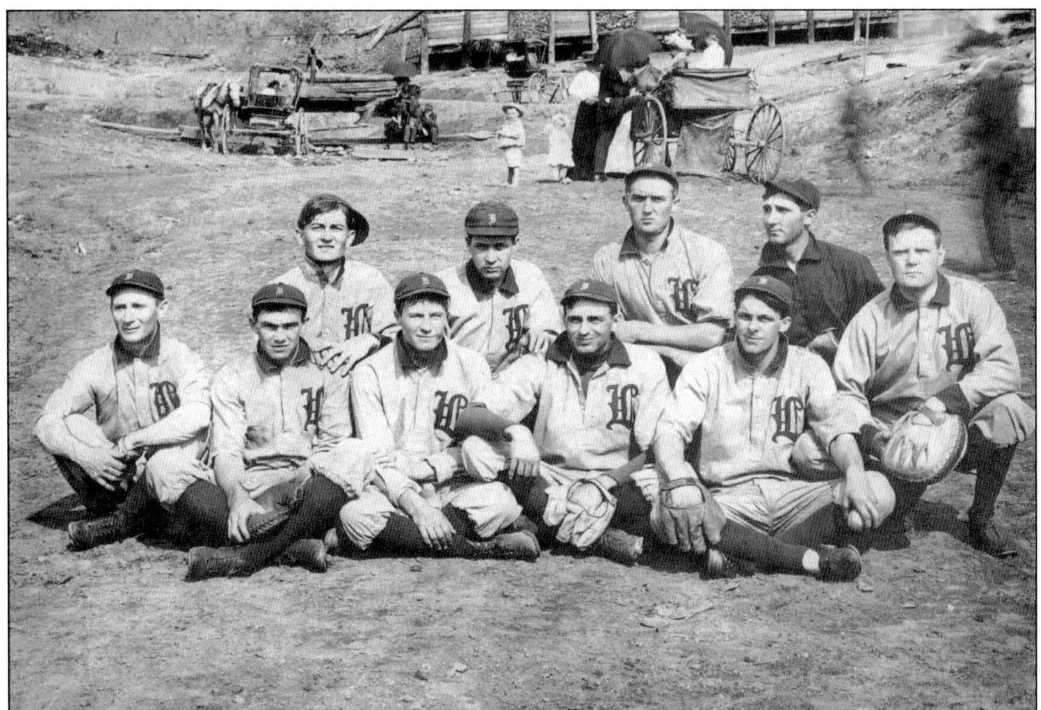

In the second row of this photograph of the 1910 Ducktown baseball team, the first player on the left is Carl Abernathy, the third player from the left is John B. Frye, and second from the right, in the dark shirt, is Ed Carver. Historical newspaper accounts suggest that the photograph may have been taken at the site of the present-day ball field in Ducktown. (Courtesy of Doris Quintrell Abernathy.)

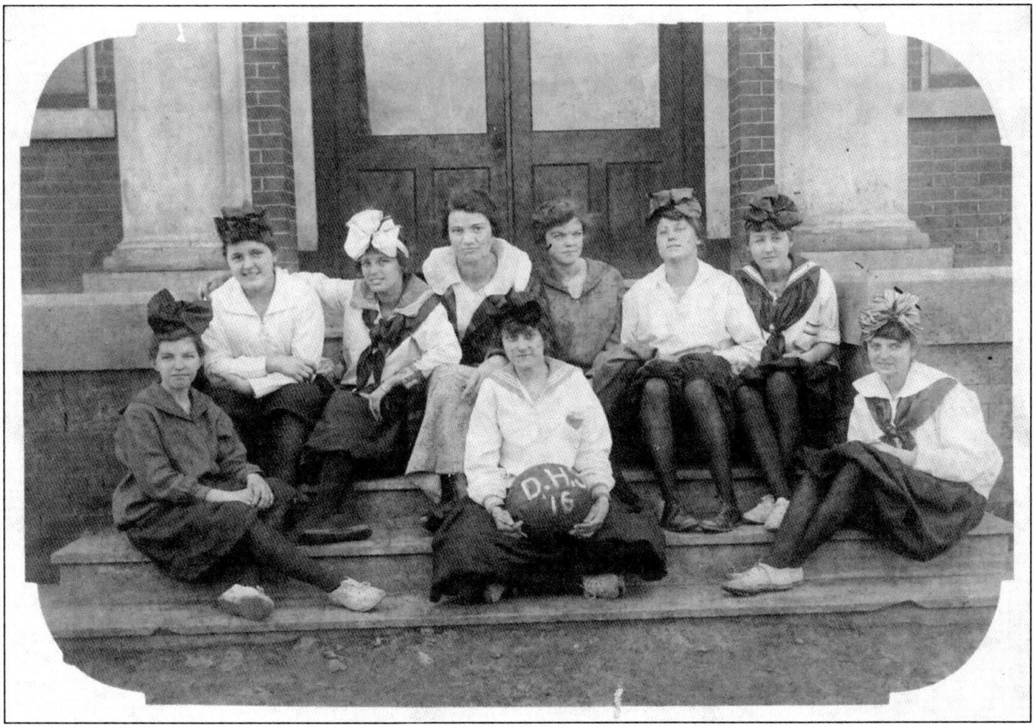

In this photograph of the 1916 girls' basketball team at Ducktown High School, Pauline (Brooks) Clay is at far left in the first row, and Edith Marchant is at far right. In the second row, from left to right, are Vera Spargo, Lucille Marchant, unidentified (probably the coach), Eula May (Brooks) Clay, Gaynelle (Bell) Hyde, and Lena Harris. The girl holding the ball is also unidentified.

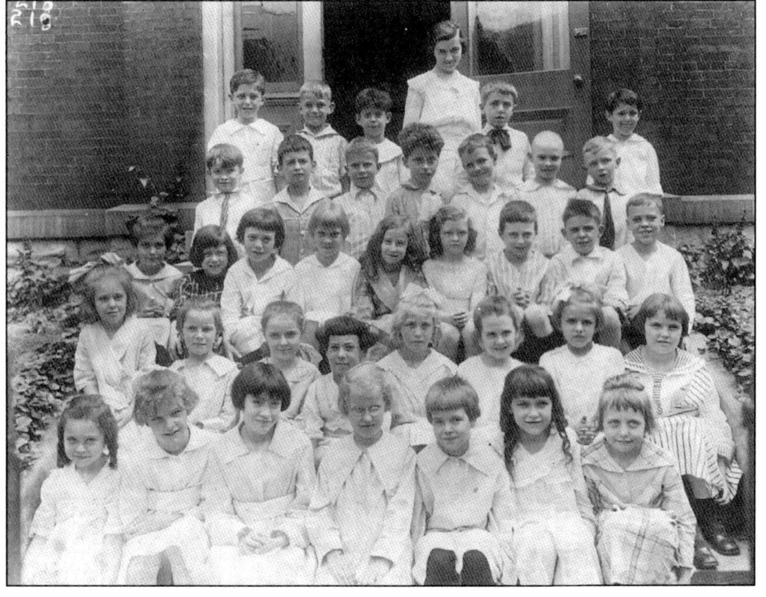

The brick grammar school at Ducktown, which sat on a hilltop beside the previously constructed high school building, was opened in January 1918. The only child who can be positively identified in this photograph is Evelyn Payne, fourth from left in the first row, who was born around 1910 and later became a teacher at Ducktown. The photograph was probably taken between 1918 and 1920.

In this 1916 photograph of a Ducktown basketball team, the player holding the ball is W.C. "Chubby" Smith. Although this group is identified on the original photograph as a Ducktown High School team, the lettering on the ball says "YMCA." The *Polk County Republican* regularly reported on games played by the YMCA baseball team during this period, but there was no mention of a basketball team.

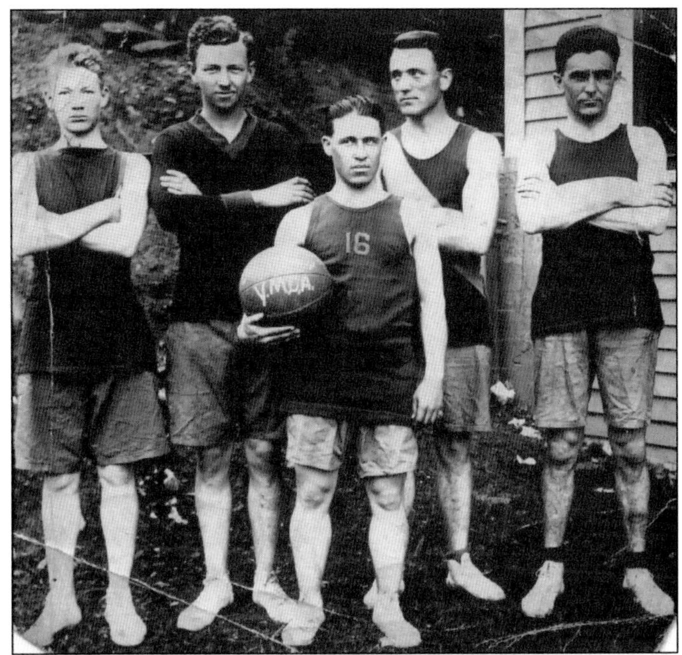

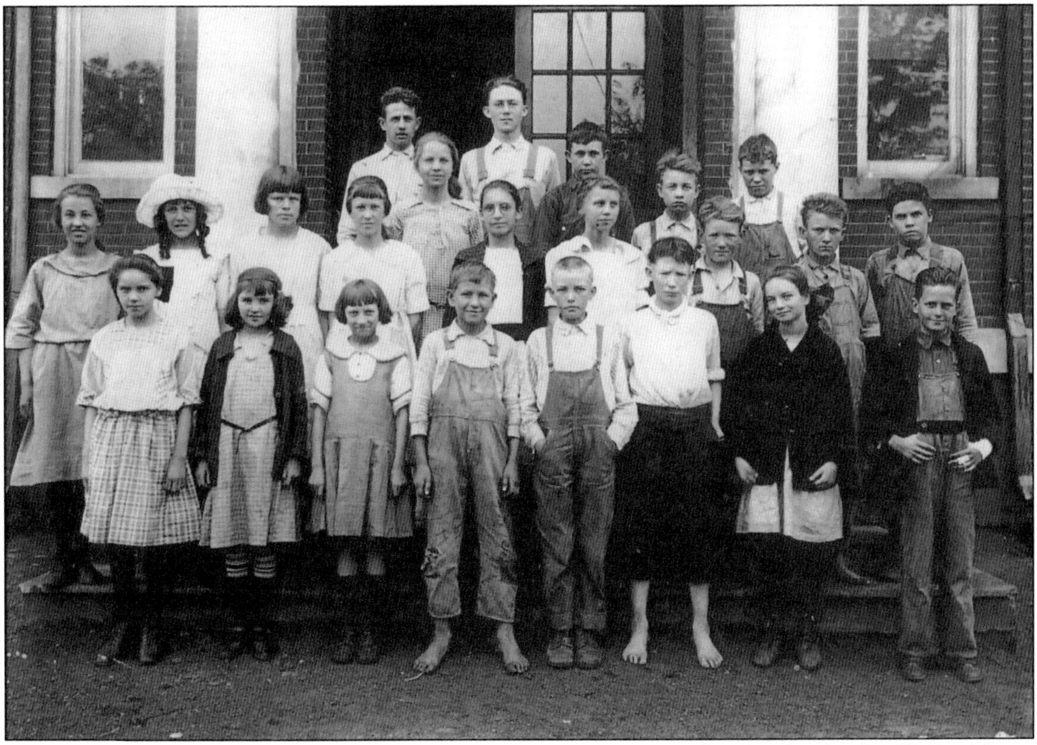

In this 1923 photograph of the fifth-grade class at Ducktown School, Claudia (Hall) Beckler is third from left in the first row. The others are Parks Simmons, Walter Hedden, Tommy Addington, Ruth Cole, Lloyd Pittman, Earl Smith, Hazel Taylor, Myrtle Hedden, Merle Belle, Cora Ingle, Gladys Rogers, Ruby Weaver, Taft Maughan, Ernest Craig, Wayne Berger, Evon Ellis, Betty Smith, Forrest Hall, Glenn Maughan, Percy Hensley, Vernelle Reed, and James Center.

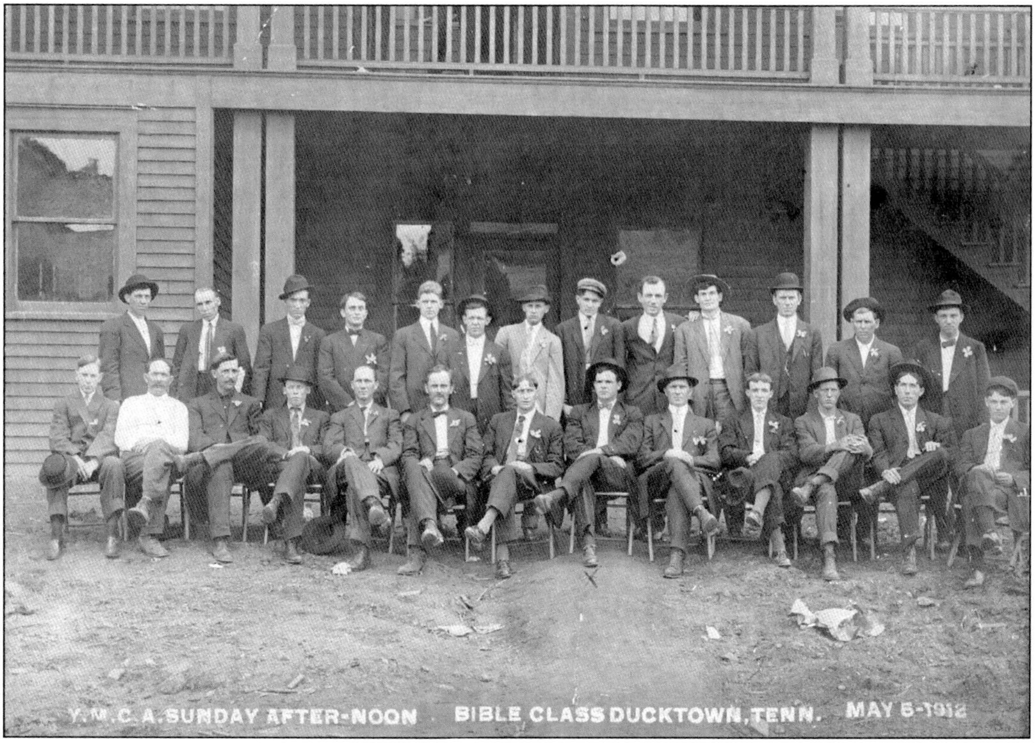

Y.M.C.A. SUNDAY AFTER-NOON BIBLE CLASS DUCKTOWN, TENN. MAY 5-1912

The YMCA building at Ducktown, constructed by TCC primarily for the benefit of employees, was dedicated in March 1911. In the era before the proliferation of automobiles, it served as Ducktown's social center, and it continued to house recreational facilities for Ducktown residents for many years afterward. The building, which burned down in 1998, stood on the southeast corner of the intersection of Highway 68 and Main Street. Above, on May 5, 1912, a Sunday afternoon men's Bible class gathers in front of the building for a formal photograph. (Below, courtesy of Greg Crawford.)

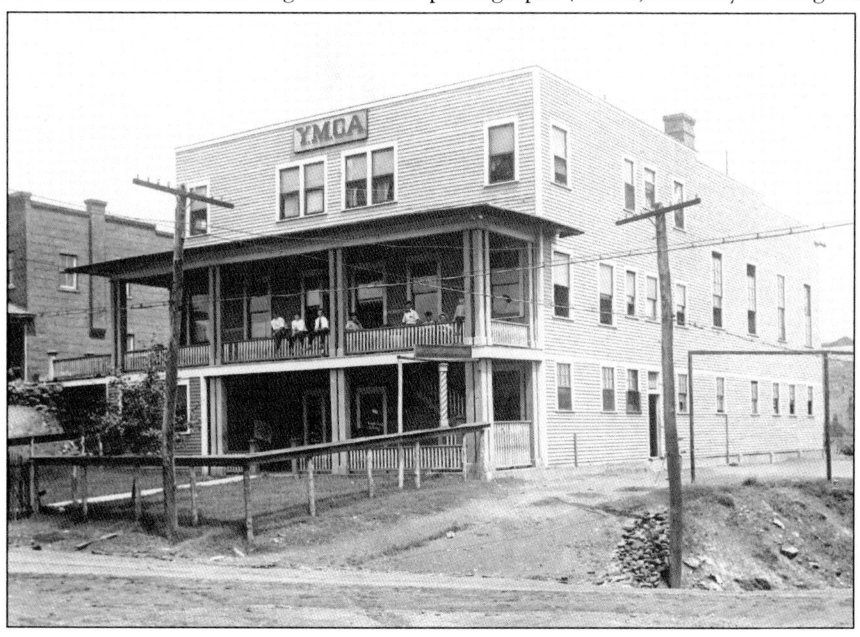

Carl Abernathy, son of Ducktown furniture dealer and funeral director George W. Abernathy, was born in 1890 and was probably around six years old in this photograph. As a young man, he moved to Copperhill and went into business for himself, first as a merchant and then as an insurance agent. He also served two terms as mayor of Copperhill. Until his death in 1979, he was one of Copperhill's best-known and best-loved citizens. (Courtesy of Doris Quintrell Abernathy.)

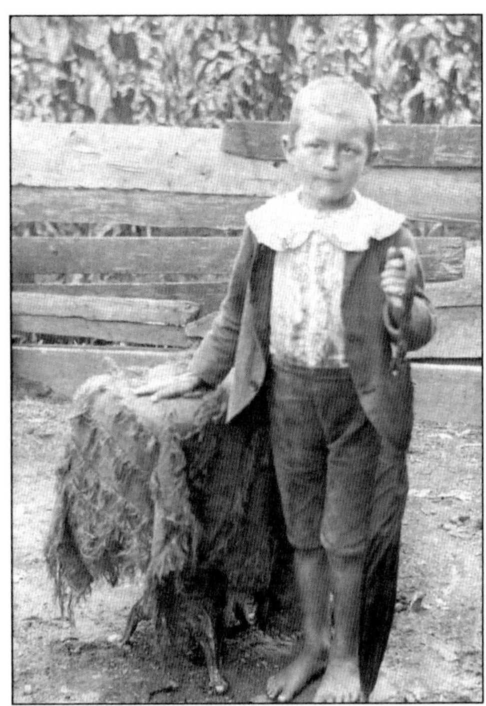

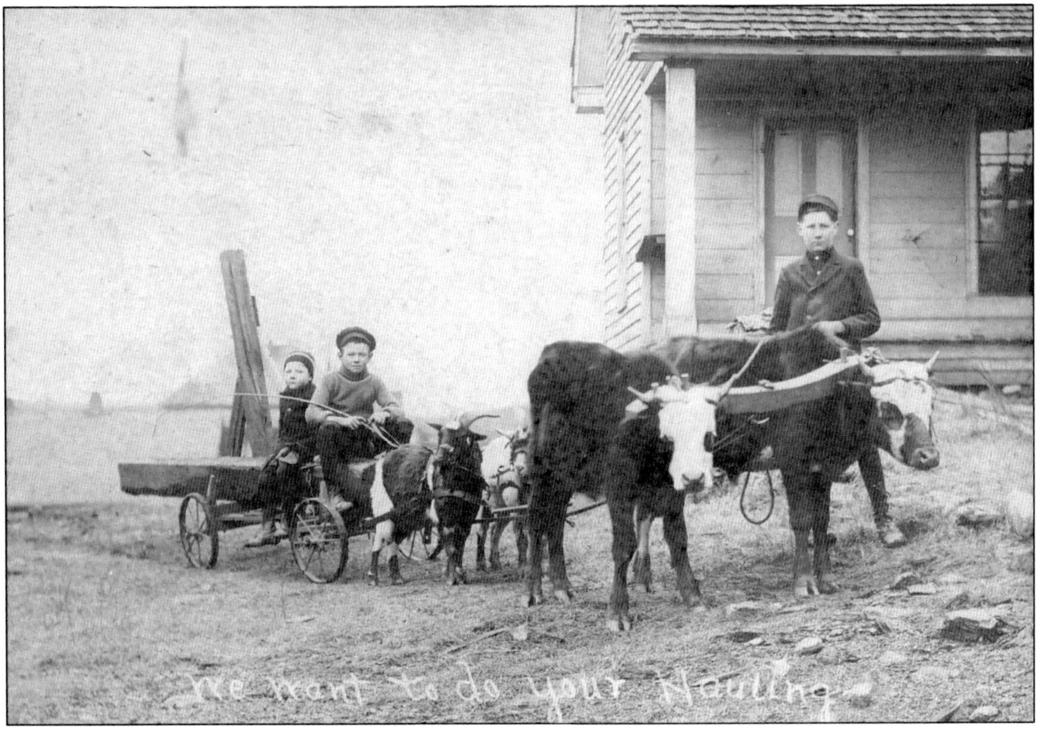

In the early 1900s, three enterprising Ducktown youngsters set themselves up in the hauling business. Equipped with a wagon, a team of oxen, and two billy goats, they advertised that they would haul anything, anywhere, anytime. In this photograph, Milton Kilpatrick holds the rope attached to the oxen, while Deyo Taylor and his younger brother Paul keep a tight rein on the goats.

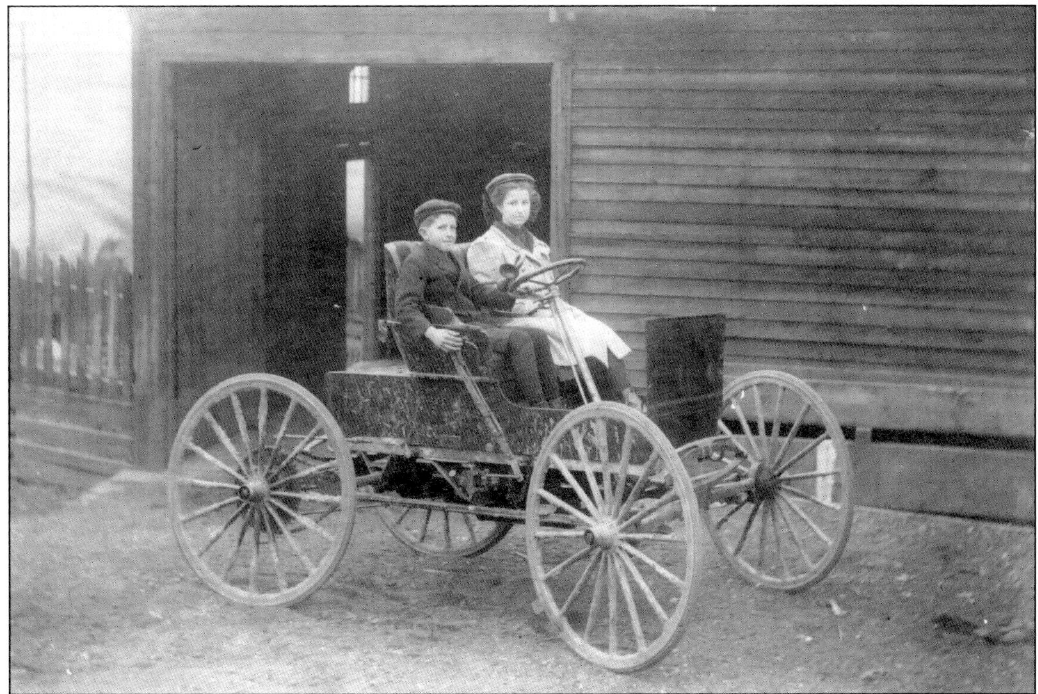

"It was on Main Street," wrote historian R.E. Barclay, "that the first automobile, a Black Motor Buggy, was seen in the Basin. Dr. L.E. Kimsey presented this rakish vehicle to incredulous Ducktown eyes in 1906." In the photograph above, Dr. Kimsey's children Parks and Pauline (later Mrs. Hoyt Campbell) pose for the camera in their father's new toy. Below, Dr. Kimsey and his wife, Helen, pose in front of their house on Main Street in Ducktown, as Pauline, Parks, and an unidentified man sit on the porch. The sign for Dr. Kimsey's medical office, which was in his home, can be seen in the window at left. (Above, courtesy of the Campbell family.)

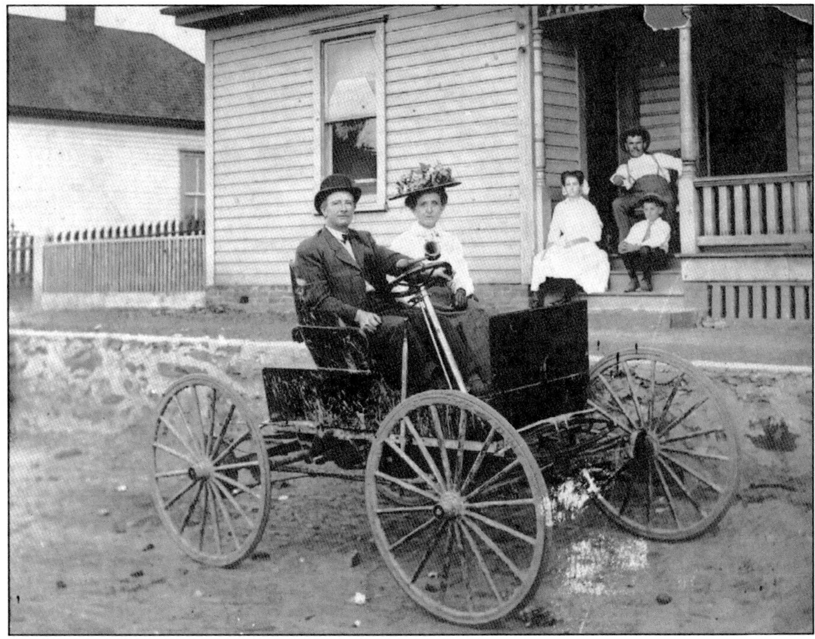

This photograph of cousins A.J. Guinn (left) and Windom Kimsey may have been taken while both were still in medical school in Atlanta. Dr. Guinn set up his practice in Ducktown in late 1915; Dr. Kimsey, son of Dr. L.E. Kimsey, apparently went into practice around the same time but died in Ducktown in 1924. In 1925, Dr. Guinn and his uncle Dr. Fred M. Kimsey, brother of Dr. L.E. Kimsey, founded what was to become the Kimsey-Guinn Hospital in Ducktown. Both men were the grandsons of Ducktown blacksmith H.P. Kimsey (see page 8).

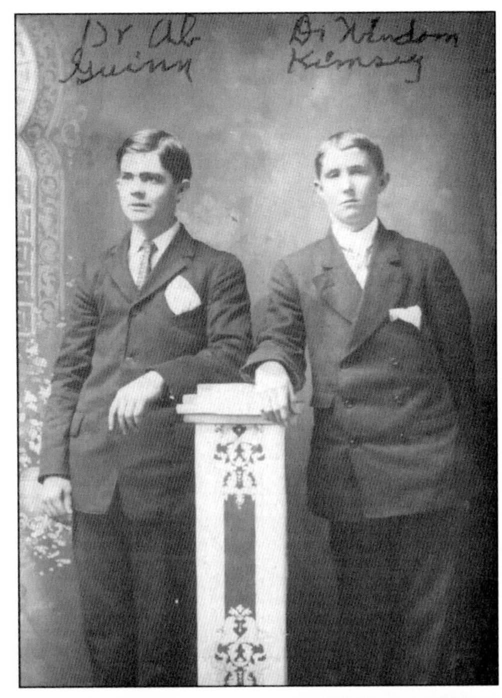

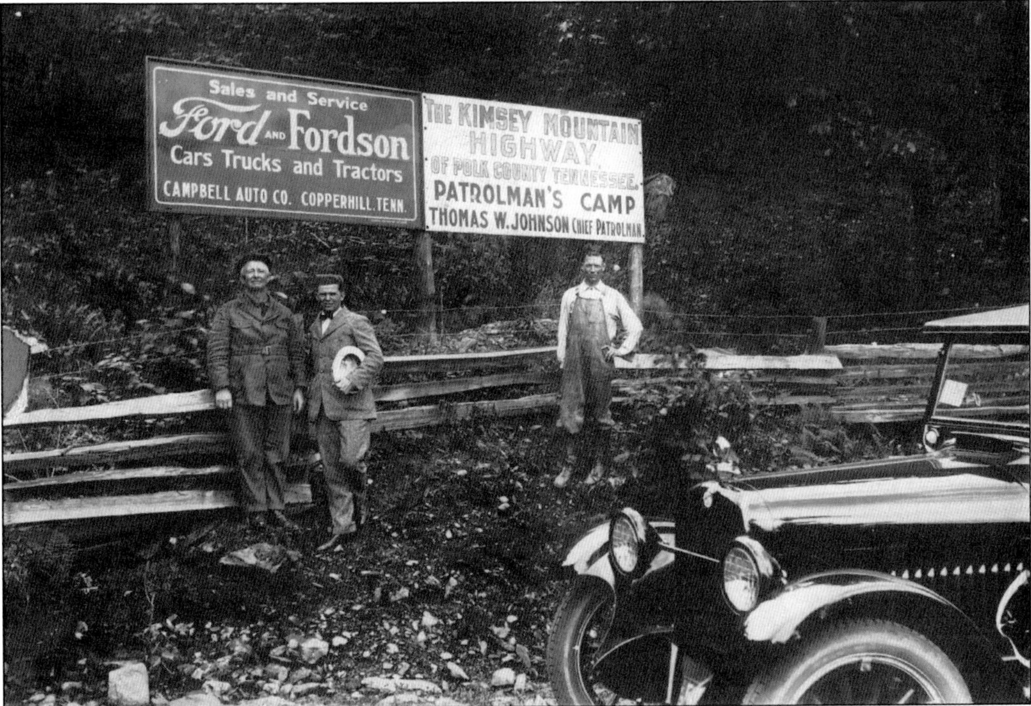

The Kimsey Highway, completed in 1920, twisted its way across Little Frog Mountain between Harbuck and Reliance. Until the Old Copper Road was rebuilt in 1931, it was the main east-west corridor for automobile traffic in and out of the Basin. In this 1920 photograph, Dr. L.E. Kimsey, principal promoter of the new road, is on the left, and chief patrolman Thomas W. Johnson is on the right.

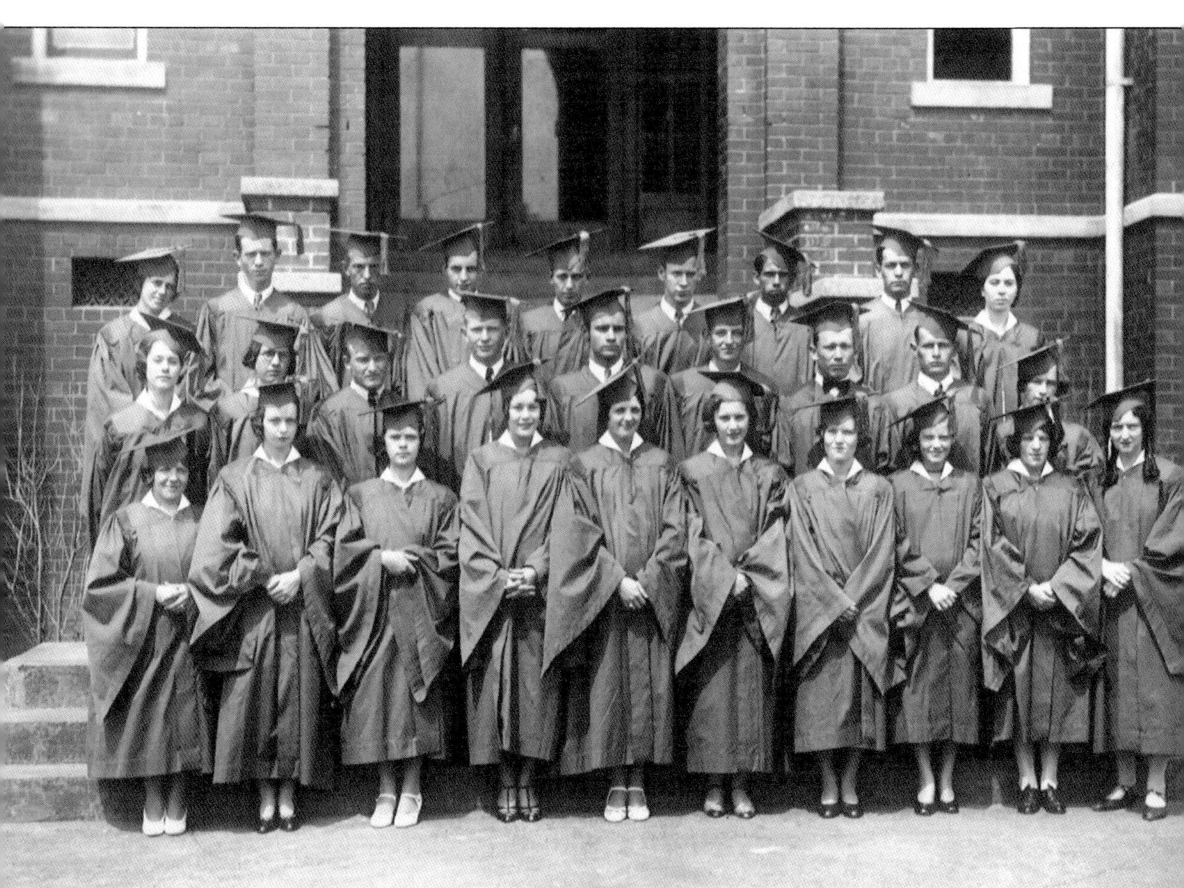

The Copperhill High School (CHS) class of 1932 poses for its graduation photograph on the front steps of the school. From left to right are (first row) Ethel Ratcliff, Ruby Woody, Leon Bailey, Jewell Woody, Lillian Clement, Elizabeth Chastain, Ruth Ledford, Mary Lois Smith, Susie Butt, and Audis Mull; (second row) Jean Johnson, Frances McGee, Ralph Chapman, J.T. "Tuck" Fry, Earnest White, Albert Sullivan, M.L. Lewis, Earl Tipton, and Verna Fair; (third row) Rheta Mae Weiss, James Panter, Hoyle Duncan, C.W. Gilliam, Lane McGee, Louis Akin, Ray Burnette, Kenton Lyle, and Blanche Tallent. There are at least three future married couples in the photograph: Leon Bailey and Hoyle Duncan, Jean Johnson and J.T. Fry, and Jewell Woody and James Panter. Among the numerous graduates who remained in the Basin were Louis Akin, who worked at the Copperhill Post Office; C.W. Gilliam, whose dental offices were on Ocoee Street in Copperhill; Jean (Johnson) Fry, who became a high school teacher; and Susie (Butt) Barrett, a familiar figure for many years at the Smelter Store. (Author's collection.)

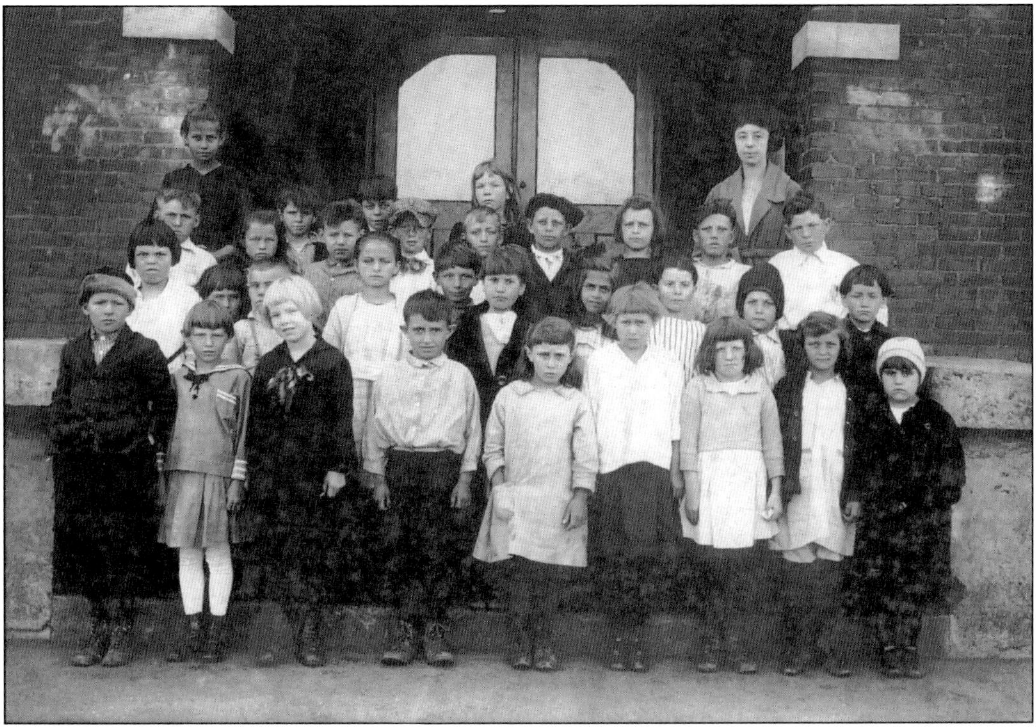

Katharine McCay (later Mrs. Roy Kincaid) is third from left in the first row of this 1923 Copperhill School class, and Susie Butt (later Mrs. Milton Barrett) is third from right. Based on the c. 1913 birthdates of these two identifiable children, this was probably a fourth-grade class. One girl in the first row defied convention by wearing white stockings instead of black. (Author's collection.)

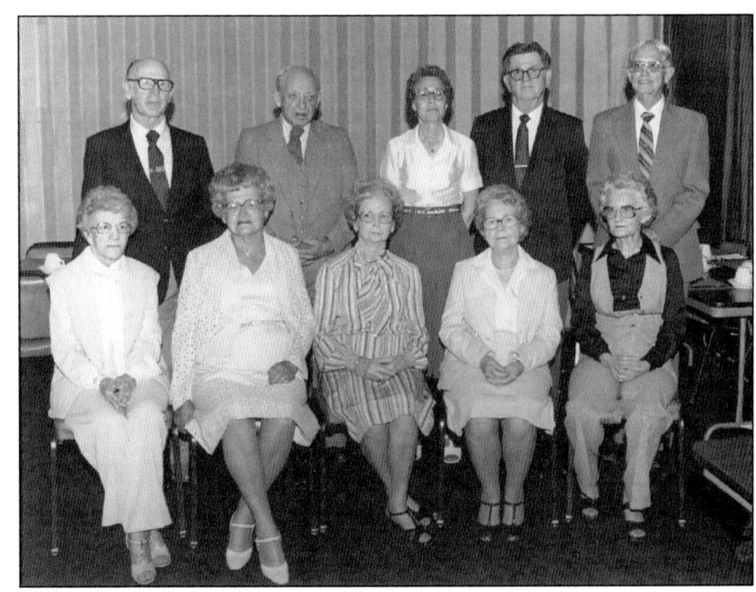

On May 15, 1982, members of the CHS class of 1932 held their 50th reunion. Those present were, from left to right, (first row) Susie (Butt) Barrett, Ruby Wilson, unidentified, Leon (Bailey) Duncan, and Elizabeth Panter; (second row) Louis Akin, Ralph Chapman, Jewell (Woody) Panter, James Panter, and Hoyle Duncan. (Author's collection.)

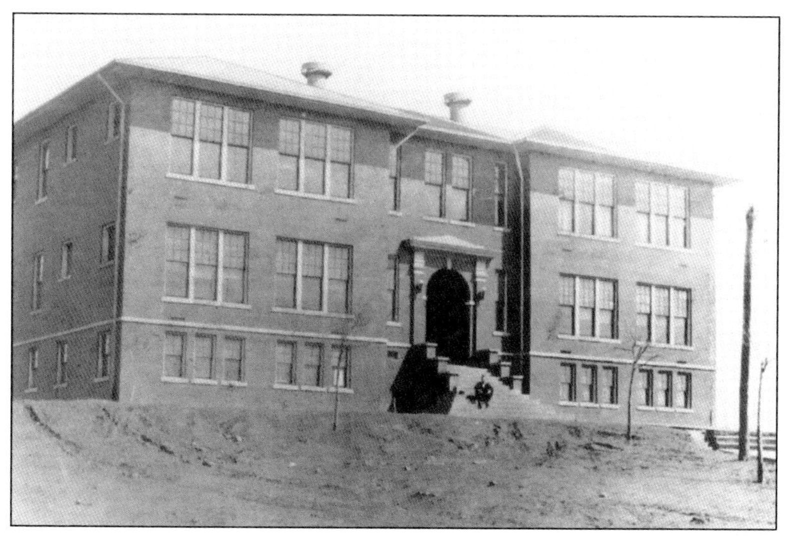

When the one-room school at Copperhill was destroyed by fire in 1910, it was replaced by a four-room brick grammar school. High school classes were temporarily held in a downtown building. In 1915, this three-story brick high school was erected on a hill above the grammar school. (Author's collection.)

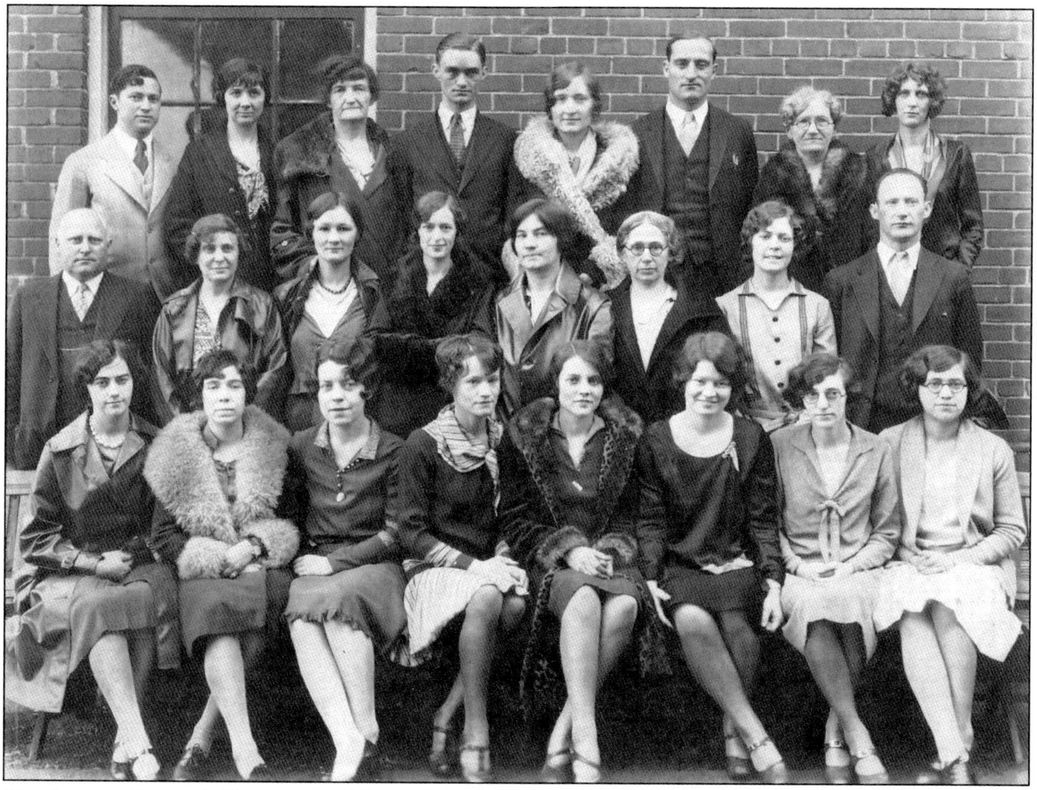

Teachers at Copperhill around 1930 are, from left to right, (first row) Helen Arp, Mildred McClain, ? Carpenter, two unidentified, Sudie Smith, ? Weaver, and Madeline Walsh; (second row) principal I.T. Sliger, Oina Mitchell, Thelma (Casteel) Richards, Ellen (Center) Ballew, unidentified, Bessie Akin, Ida Belle (Thomason) Earnest, and Fred German; (third row) Hugh Yokum, unidentified, Mary Cook, Emmett Weeks, Mary Penn, coach E.A. Hetzner, Sallie Cook, and unidentified. (Courtesy of Peggy Walter Kilpatrick.)

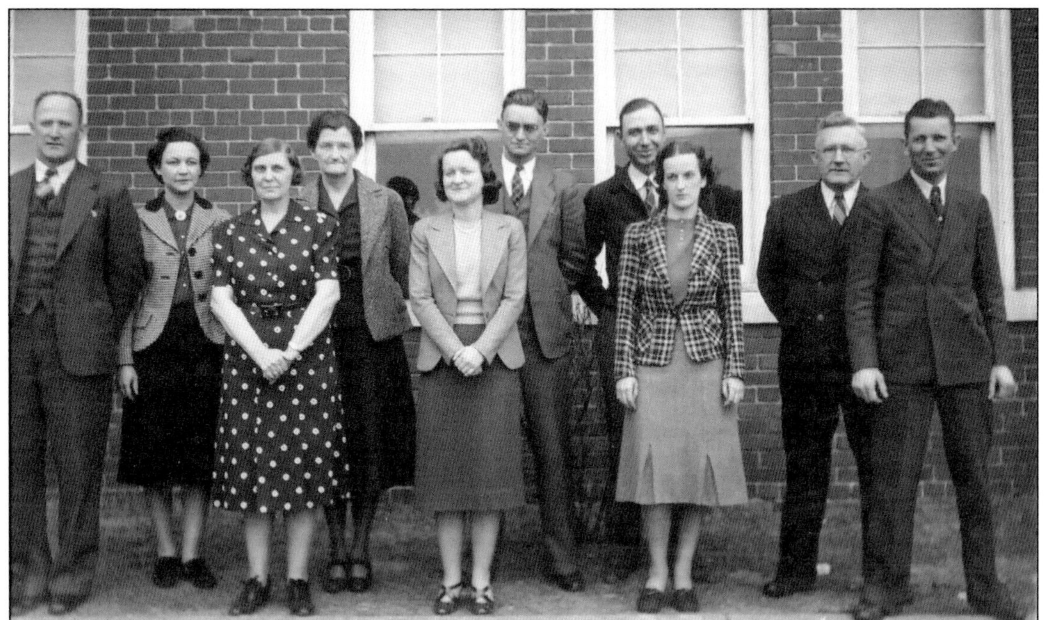

Members of the teaching staff at Copperhill High School in 1939 are, from left to right, Fred German (science), Marguerite Kollock (home economics), Oina Mitchell (music), Mary Cook (English), Marie Graham (language), Emmett D. Weeks (mathematics), Howard Garbee (vocational), Martha Porter (librarian), principal J.M. Reedy, and coach W.D. Sneed. (Author's collection.)

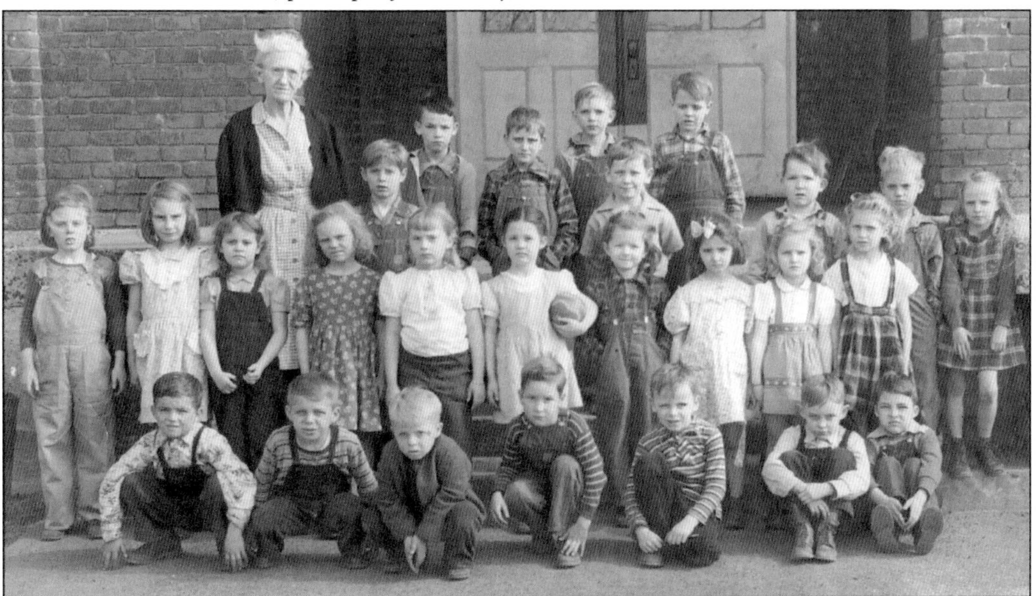

Students in the much-loved "Miss Sallie" Cook's 1947–1948 first-grade class at Copperhill School are, from left to right, (first row) Johnny Davis, Jimmy Ratcliff, Gilbert Ratcliff, Eddie Wilson, Billy Dalton, James Newell Smith, and Wayne Paul Jones; (second row) Frieda McAllister, Dorothy Nix, Jean Parris, Alice McGill, Janice Duncan, Sherry Rogers, Glenda Sue Panter, Christine Jones, Glenda Tanner, and Georgia May Wilson; (third row) Cook, Max Hughes, Billy Brackett, Charlie Messer, Leonard Galloway, and Carolyn Rembert; (fourth row) Wendell Rembert, Clint Martin, Simond Patterson, and Earnest Hammock. (Courtesy of Wayne Paul Jones.)

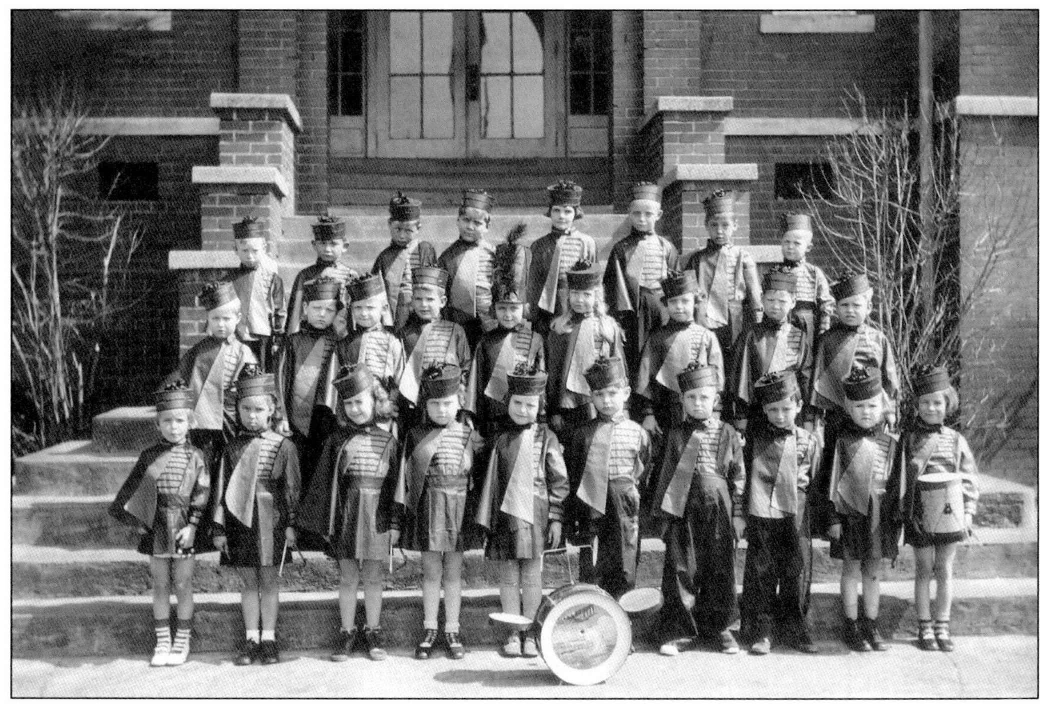

In the many years of Oina Mitchell's tenure, musical education was an essential part of the curriculum at Copperhill School. In the 1938–1939 school year, the school boasted both a rhythm band, above, and a toy orchestra, below. Mitchell also directed the high school orchestras and glee clubs, as well as the annual grammar school operettas. (Both, author's collection.)

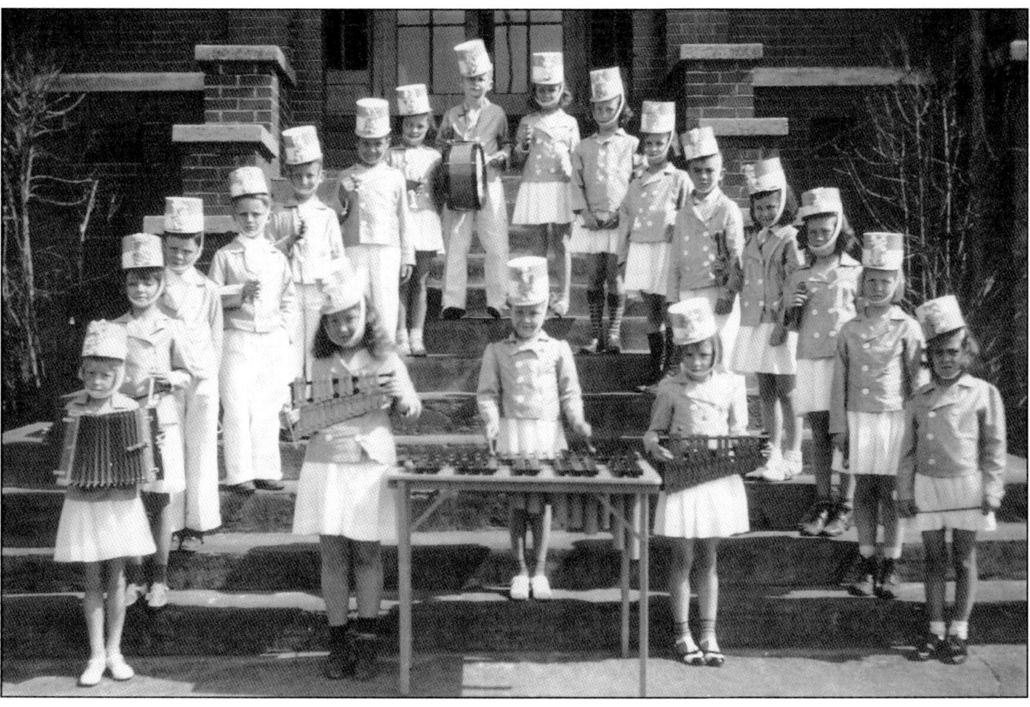

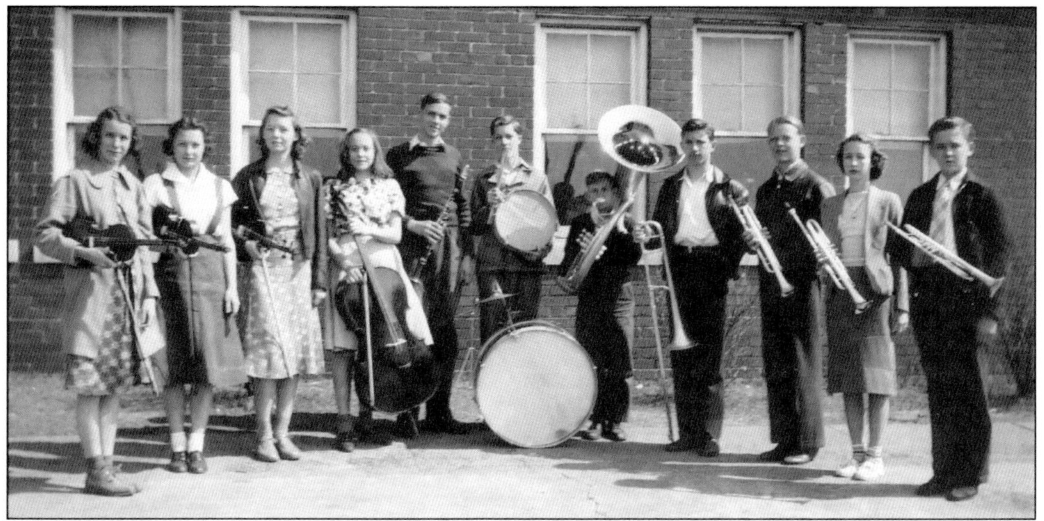

Members of the Copperhill High School orchestra for the 1938–1939 school year are, from left to right, Betty Beckler, Virginia Quinn, Evelyn Cutts, Marion Kerns, Bobby McCarter, Warren Sarrell, Dan Thompson, Eugene Key, Kathleen Harper, and Lynn Hicks. (Author's collection.)

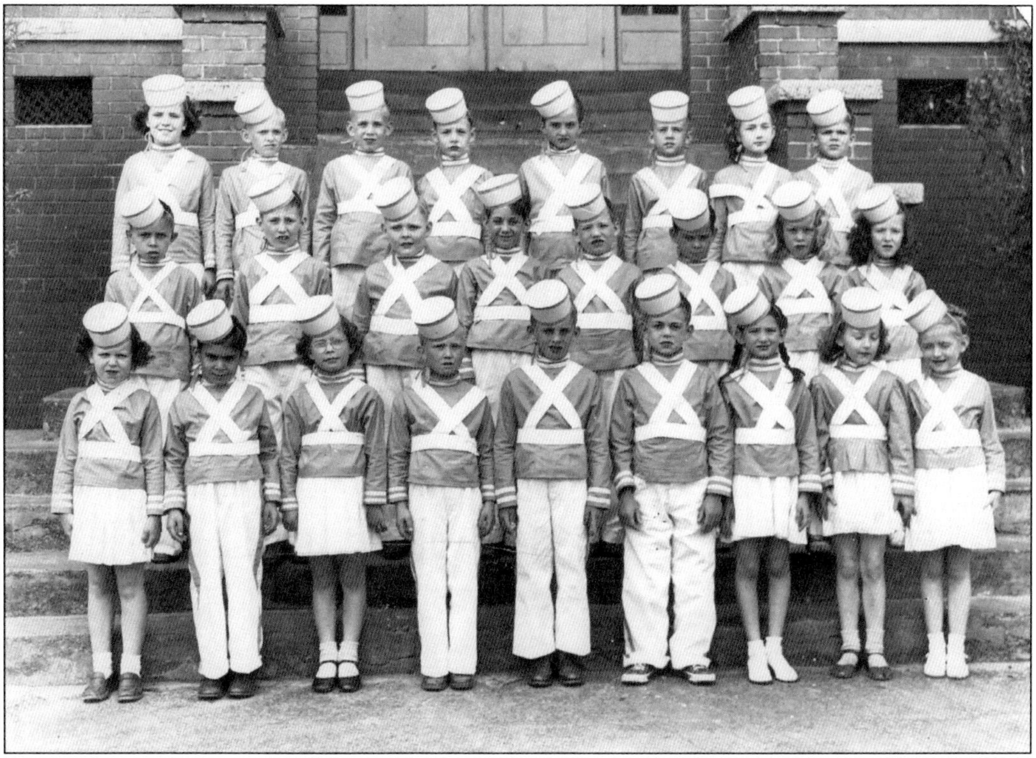

From left to right in this c. 1949 Copperhill rhythm band photograph are (first row) Margie Humphreys, George Wayne Shibley, Sue Stuart, James "Bee" Hicks, John O'Kelley, Tony Middleton, Jane Ballew, Rebecca Adams, and Barbara Self; (second row) James Mitchell, Travis Goss, Phil Hall, Ferris Maloof, George Cobb Jones, Eddie Massengale, and two unidentified; (third row) Patricia Roach, Haines Hill, Britt Burns, Billy Middleton, Jackie Russell, Edwin Light, Mary Nan Higdon, and Mike Morgan. (Courtesy of Travis Goss.)

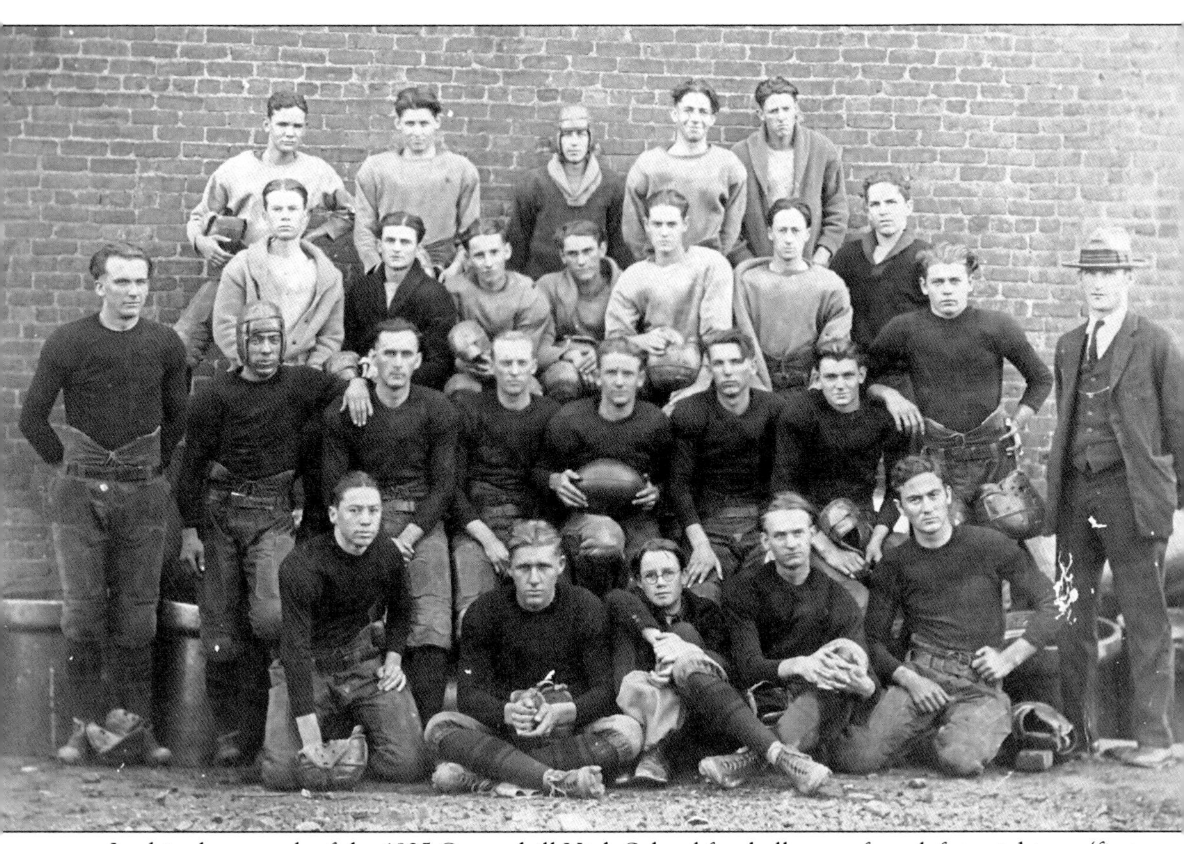

In this photograph of the 1925 Copperhill High School football team, from left to right are (first row) Leo Jones, Bethel Stover, manager J.S. Akin, Earl Arp, and Earl Huffman; (second row) Ally Johnson, Raymond "Hoot" Gibson, Earnest Johnson, Roy Chapman, Fred McCay, Charles Sheets, Reece Higdon, Frank Morgan, and coach Fred German; (third row) Clyde Ledford, Harold Beavers, Tom Adams, Clint Angel, Kermit Adams, Hugh Brodie, and Albert Lewis; (fourth row) Cecil Dalton, Glenn Gardner, Lake McGee, Otis "Toby" Gibson, and Earl Gardner. This was the first known football team at CHS, and its initial season was apparently not an unqualified success. Of the four games that can be documented, the team lost 0-31 to Polk County High School and 0-18 to Alcoa High School, tied Tellico Plains High School 0-0, and eked out a 7-0 victory over Maryville High School.

In the first row of this photograph of the 1937–1938 CHS basketball team are, from left to right, Carl "Nanners" Buchanan, Mark Queen, Dwight Tipton, Perry Tipton, and Billy Brackett. At far left in the second row is Red Stover, and third from left is Howard Kaylor. The coach is Roy Ferguson.

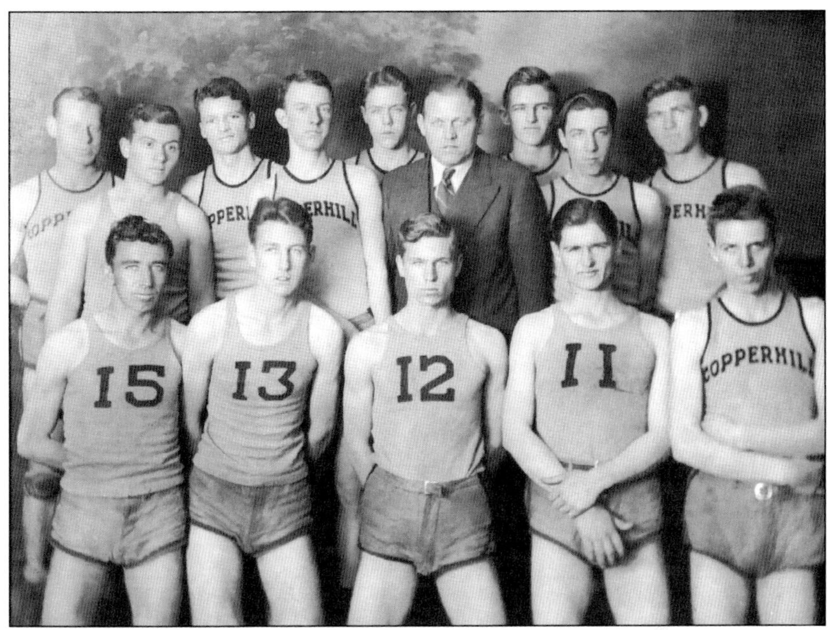

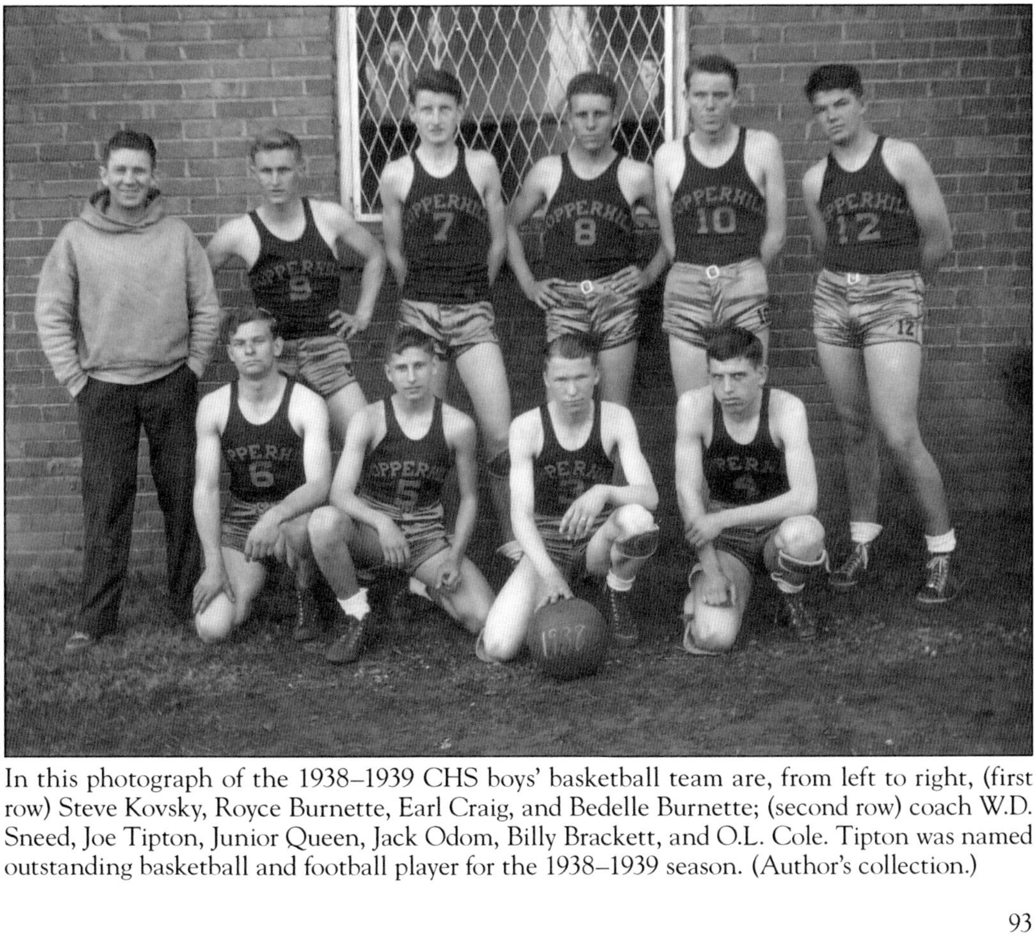

In this photograph of the 1938–1939 CHS boys' basketball team are, from left to right, (first row) Steve Kovsky, Royce Burnette, Earl Craig, and Bedelle Burnette; (second row) coach W.D. Sneed, Joe Tipton, Junior Queen, Jack Odom, Billy Brackett, and O.L. Cole. Tipton was named outstanding basketball and football player for the 1938–1939 season. (Author's collection.)

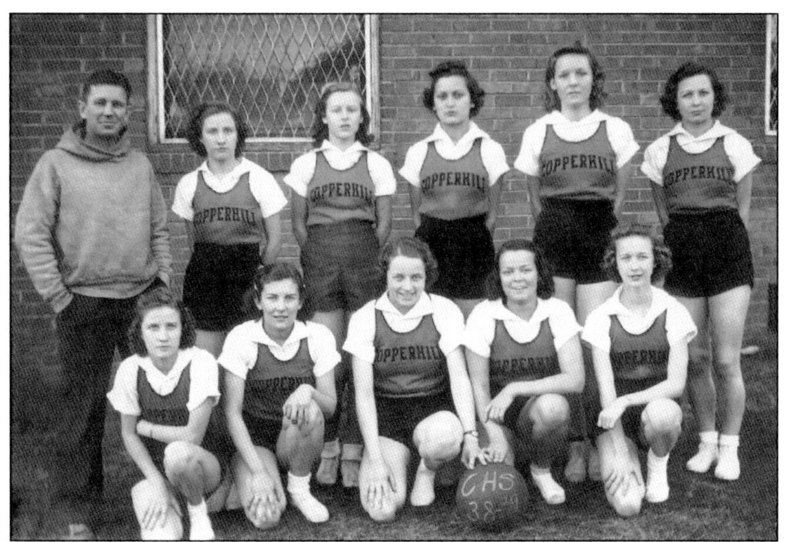

From left to right in this photograph of the 1938–1939 CHS girls' basketball team are (first row) Daisy Elrod, Evelyn Cobble, Louise Young, Hazel Cole, and Kathleen Harper; (second row) coach W.D. Sneed, Violet Elrod, Betty Gobble, Cleo Quinn, Evelyn Cutts, and Virginia Quinn. (Author's collection.)

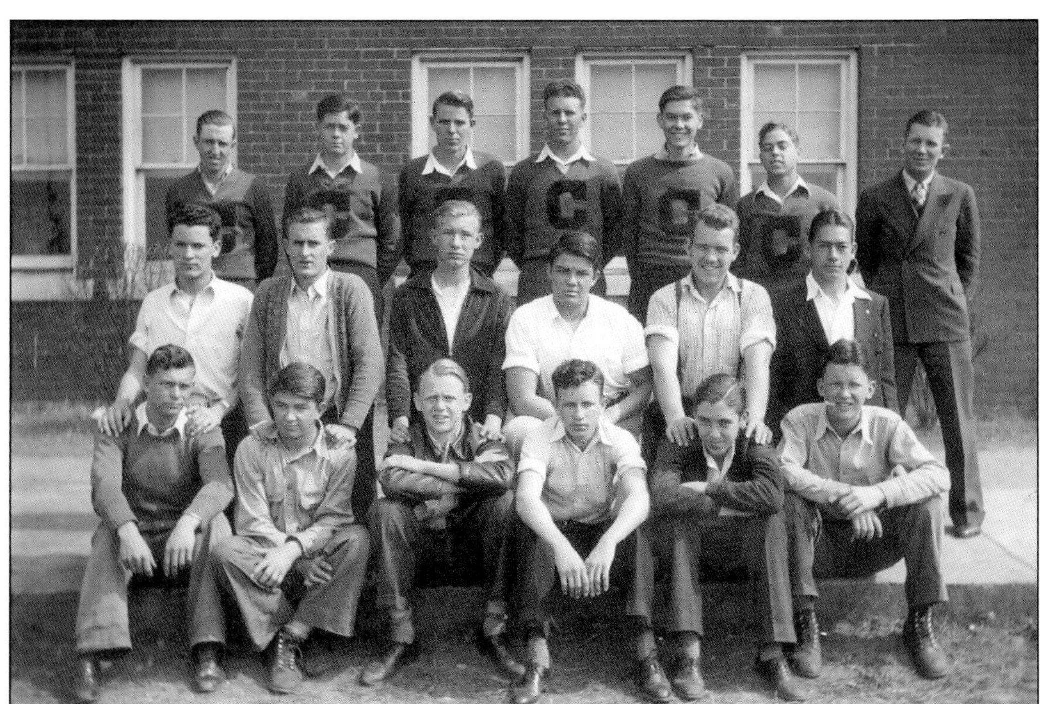

In this photograph of the 1938–1939 CHS football team, from left to right are (first row) Steve Kovsky, Charles Cole, Eugene Key, H.M. Crawford, manager Robert Weiss, and Larry Dalton; (second row) Louie Dalton, Joe Tipton, Billy Howell, O.L. Cole, A.J. Hughes, and Roland Barnes; (third row) H.C. Chapman, George Speer, Billy Brackett, Jack Odom, David McNaughton, Clement Bailey, and coach W.D. Sneed. (Author's collection.)

In this 1957 photograph, majorettes at Copper Basin High School pose with band director Boyd Williams. From left to right are (first row) Peggy Walter, June Simonds, and Beadie Mae Verner; (second row) Shelby Standridge, Sue Lewis, Boyd Williams, Ann Williams, Judy Coffey, and Patricia Powell. (Courtesy of Peggy Walter Kilpatrick.)

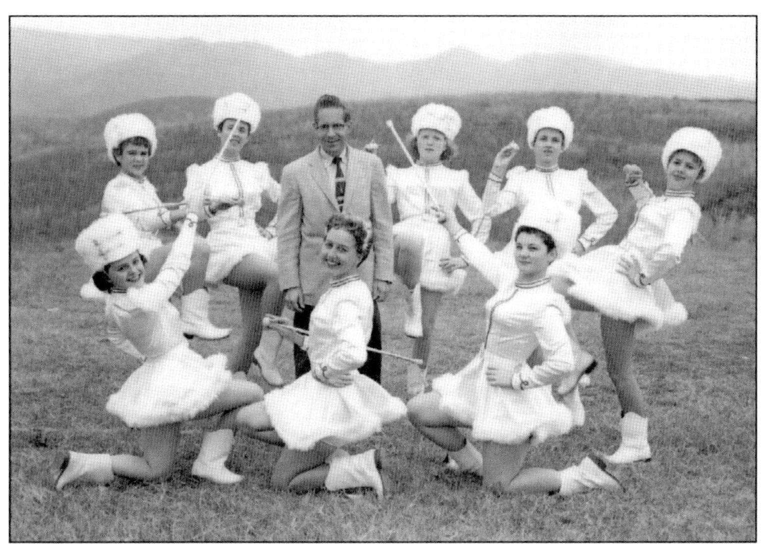

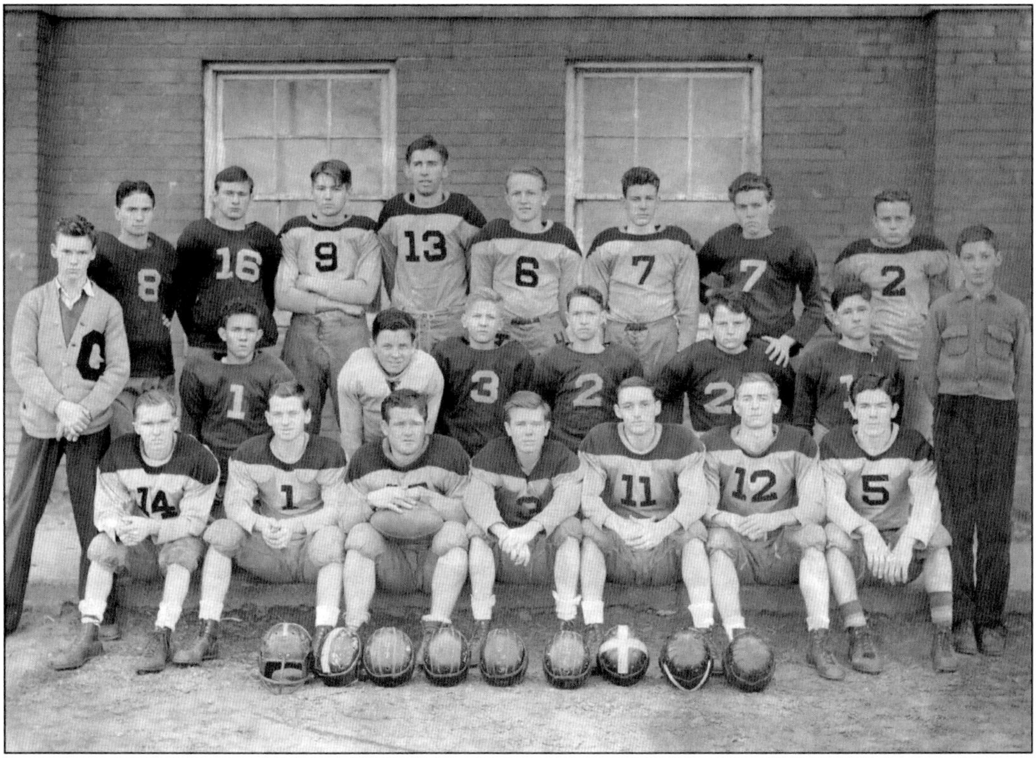

Members of the 1942–1943 CHS football team are, from left to right, (first row) Bobby Panter, Gordon Jones, German Davis, Carl Panter Jr., Charles Panter, William "Junior" Sparks, and Kent Morgan; (second row) Jack Panter, Kenneth Wilcoxon, Larry Swafford, Warren Chancey, Bobby Stanley, Paul Stepp, Charles Harrison, and Don Hetzner; (third row) Lloyd Helton, Ray Gobble, Vince Williamson, Walter Marlette, Elvin "Junior" Harper, Ray Beckler, Vernon Bidez, and Jack Terry. (Courtesy of the heirs of Polly Morgan.)

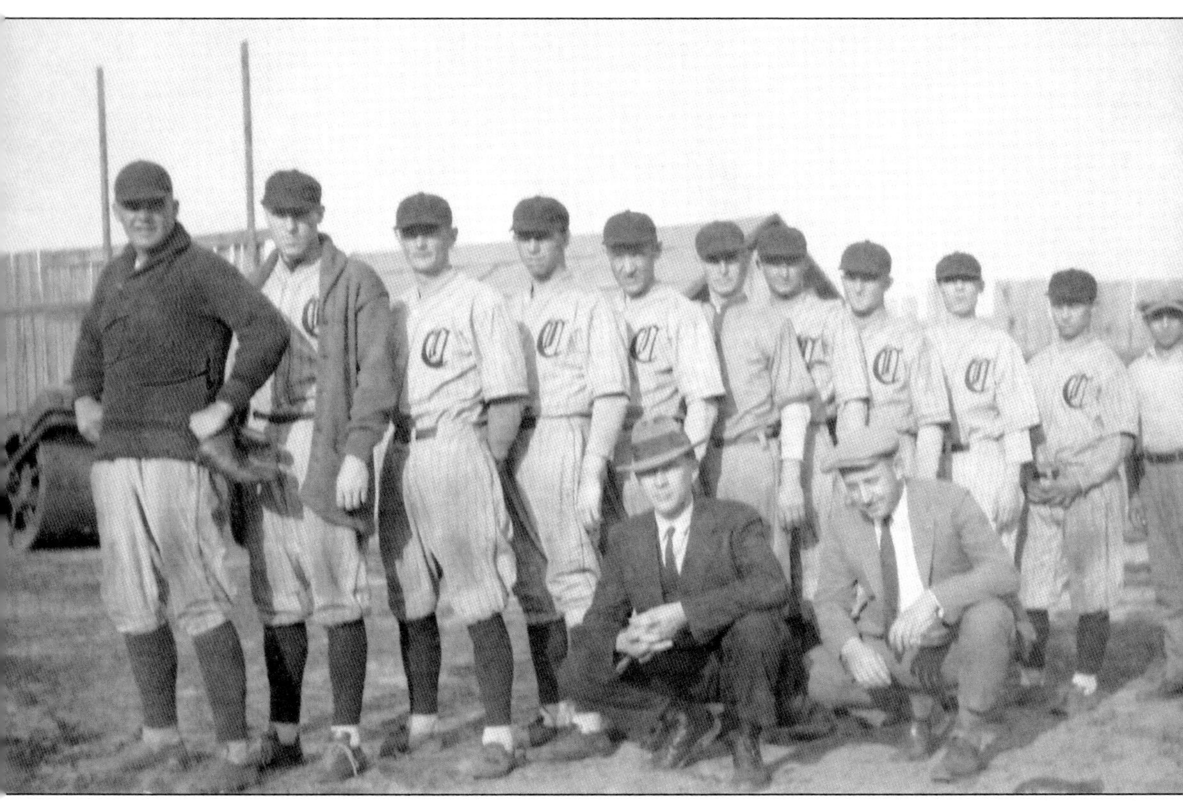

"The company supported a strong semi-professional baseball team," wrote Copperhill High School principal John Carter in his 1977 autobiography. "Most of their players were ex-college men who were given jobs with a view toward baseball ability." This photograph of an early Copperhill team, apparently taken sometime between 1924 and 1929, includes at least three players who had been recruited in this fashion: Tiny Swafford (far left), Nigal Marlette (fourth from left), and Walter Davis (unidentified). All three had played briefly with minor league teams before joining the company and putting on their Copperhill uniforms. Carter, who was principal at CHS from 1917 to 1929, is sixth from left. Others listed but unidentifiable in the photograph are Walter Smith, Lake McGee, and ? Parr. Kneeling in front are TCC company physician Dr. C.W. Strauss (left) and chief clerk Paul Norton. Basin native and major-league catcher Joe Tipton once included Tiny Swafford in his list of the three best hitters he had ever known, right behind Ted Williams at No. 1 and Joe DiMaggio at No. 2. (Author's collection.)

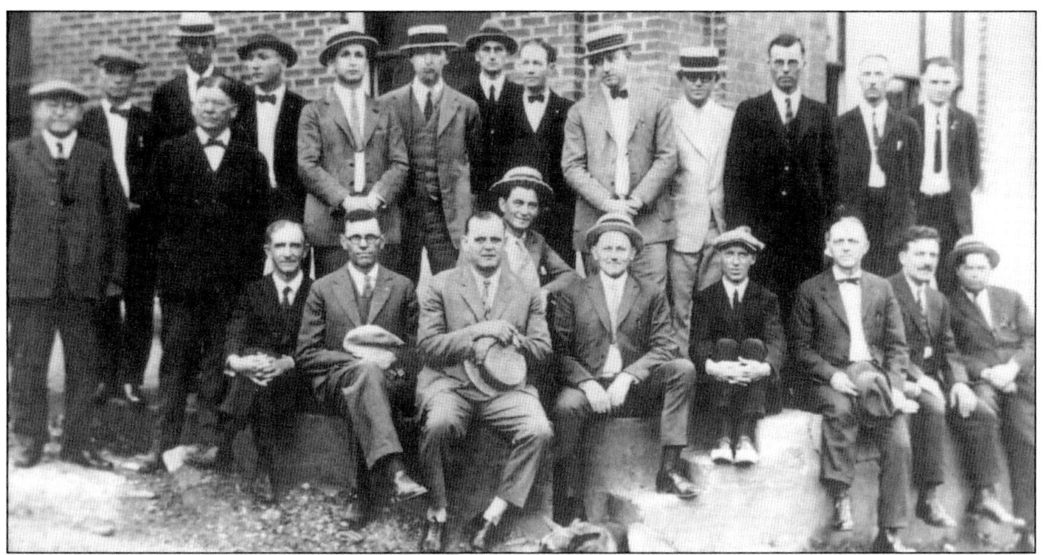

From left to right in this 1922 photograph of the Copperhill Kiwanis Club are (seated, first row) George Gilmore, Hayden Bell, Dr. W.Y. Gilliam, Carl Abernathy, Dr. Thomas J. Hicks, D. Claude McCay, W.R. Dickens, Gus Grevis, and Luther Abernathy; (standing, second row) C.L. Knoedler, R.P. "Pink" McCarter, Charles Barnes, Dr. W.L. Cutts, A.B. McCay, R.E. Barclay, Dr. C.W. Strauss, W.H. Patton, M.C. King Jr., Paul Norton, Hoyt Campbell, G. W. Kilpatrick, Rev. M.H. Rambo, and Homer Seaton.

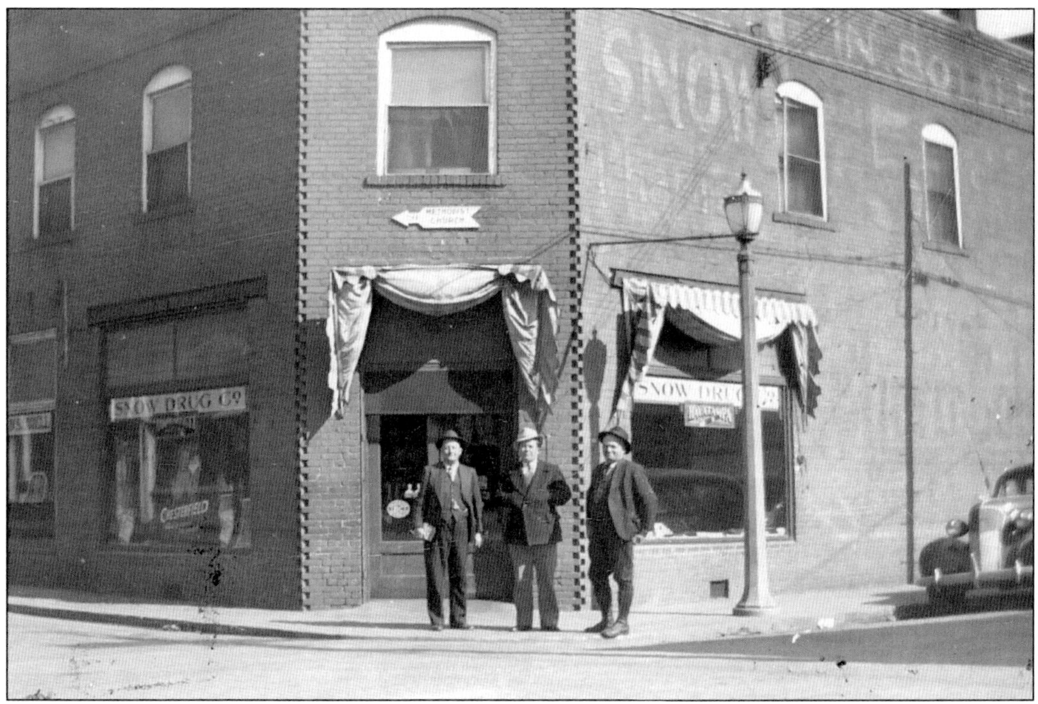

Around 1940, Copperhill physician H.P. Hyde (left) poses with Polk County deputy sheriff Frank Clayton (center) and justice of the peace Creed Brock. Pharmacist Joseph E. Snow ran the pharmacy in the background, which had formerly been the Jones Drug & Confectionery Company. (Courtesy of Peggy Walter Kilpatrick.)

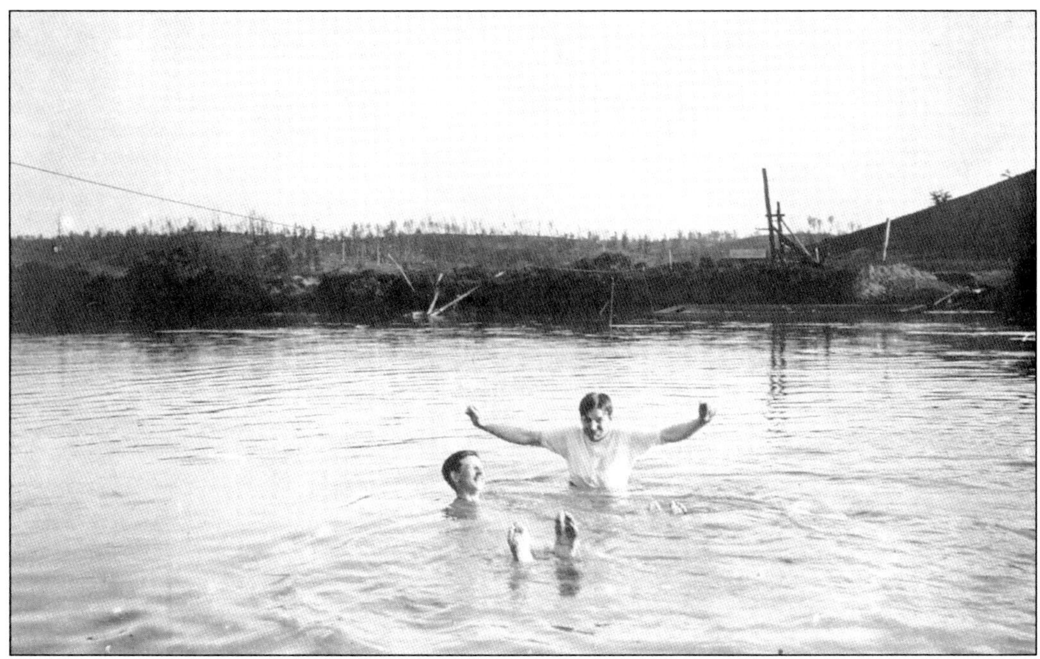

Although the building of the TCC smelter and plant facilities at McCays was serious business, the lives of Edward Oscar Boak and his fellow engineers were not all work and no play. Judging by the position of their feet, these happy swimmers in the Ocoee River around 1905 appear to be sitting on the bottom of the river.

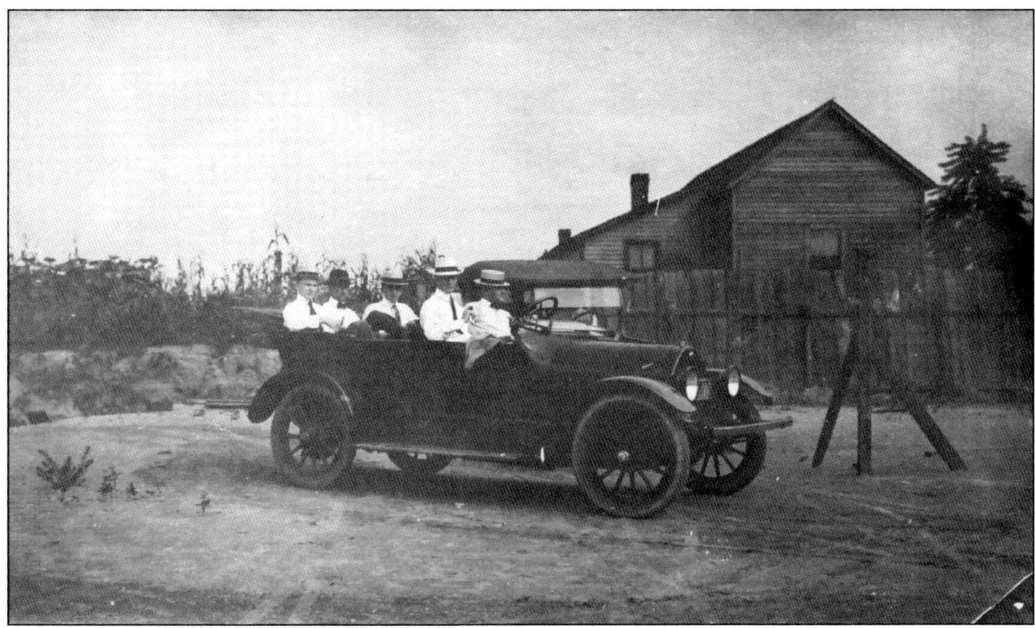

Milton A. Caine is at the wheel in this photograph, probably taken around the time of his promotion to TCC assistant manager in 1916. During this period, the company had no resident general manager; Caine was responsible for local management of TCC until he was transferred to the New York office in 1925 and succeeded by F.J. Longworth.

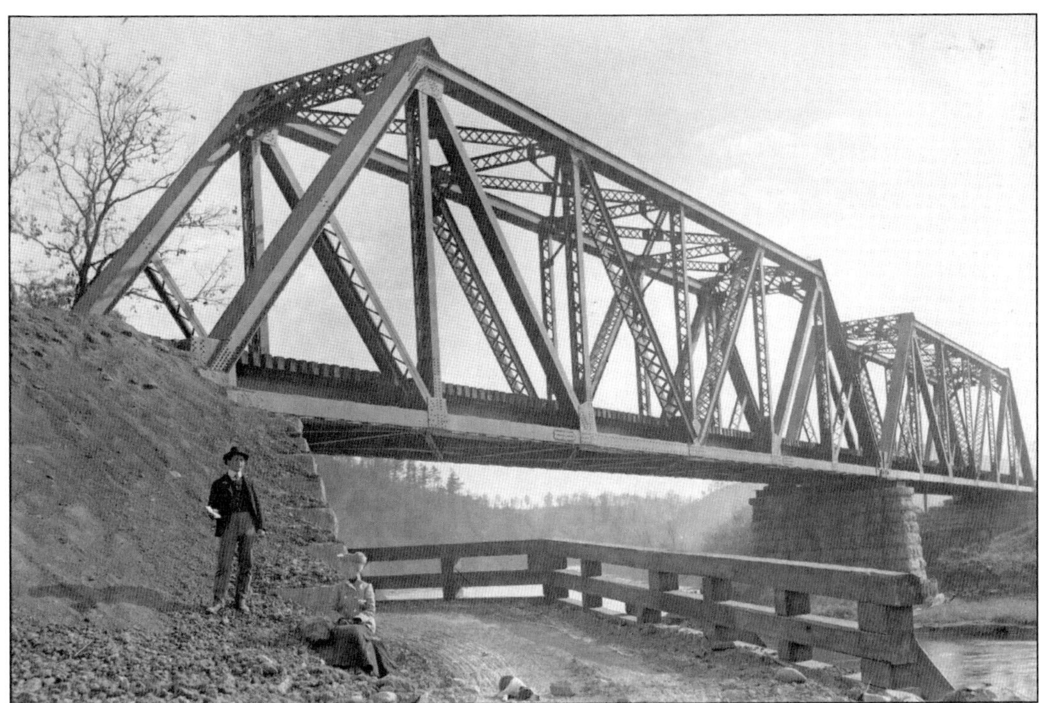

The railroad bridge across the Toccoa River near McCays (above) was built by the Marietta & North Georgia Railroad in 1889 and acquired by the L&N Railroad in 1902. It stands about a mile and a half from Copperhill, on the Georgia side of the state line, and is essentially unchanged today. In addition to providing a vital link between the Ducktown district and the outside world, the bridge was also a scenic destination for a Sunday outing. E.H. Westlake photographed the unidentified couple (and dog) above at the bridge in 1904. Below, about 20 years later, another couple poses for a Kodak snapshot in the L&N Railroad yard at Copperhill. (Below, author's collection.)

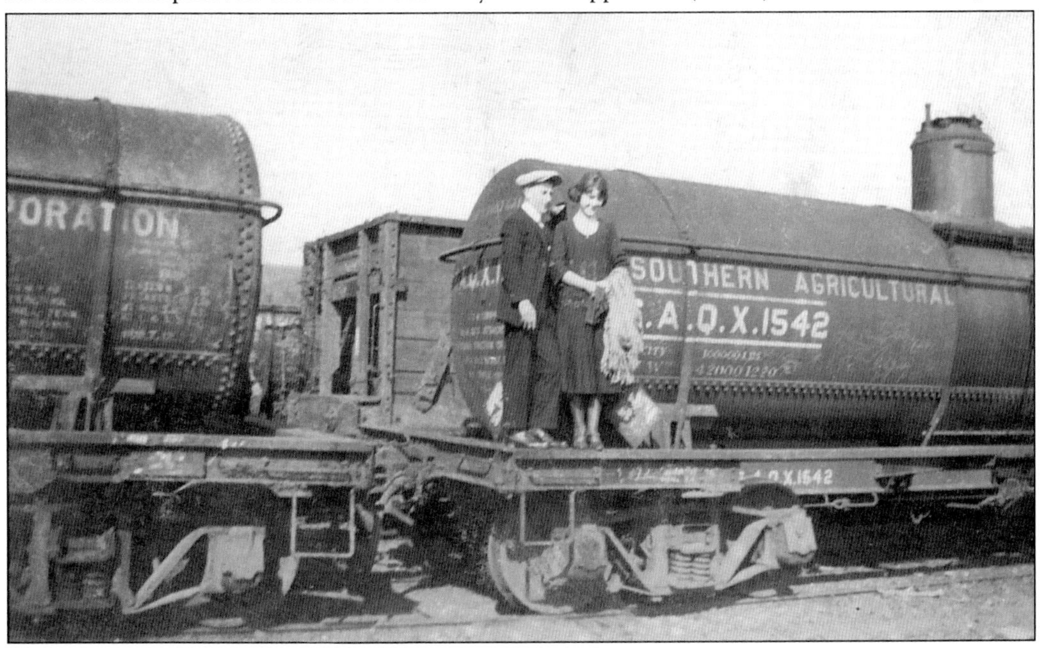

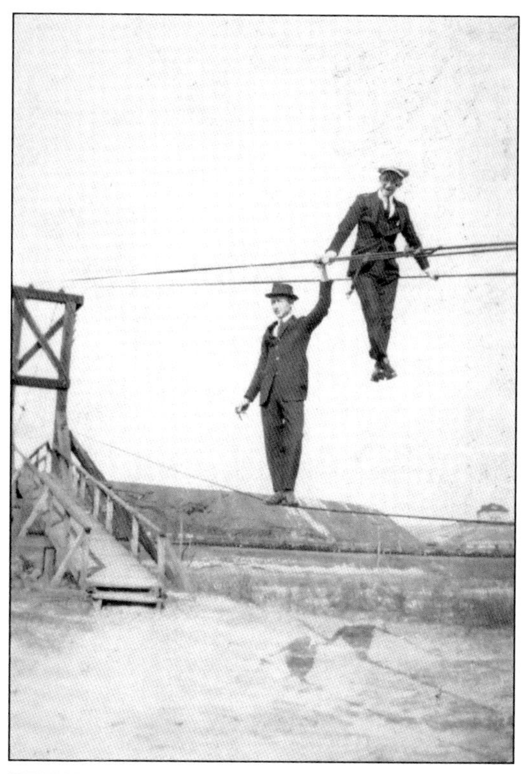

When the TCC foreign labor camp at Copperhill was built in the early 1900s, a footbridge connected the camp with the plant facilities on the other side of the Ocoee River. In the mid-1920s photograph at left, two unidentified young men show off their acrobatic skills at what appears to be the labor-camp end of the footbridge, with the slag pile from the TCC smelter in the background across the river. The building at far right may have been the TCC general office. The undated photograph below shows the site of the camp and footbridge at the mouth of Fightingtown Creek. Some local citizens remember the footbridge, which was still standing in the same location in the 1940s. (Left, author's collection.)

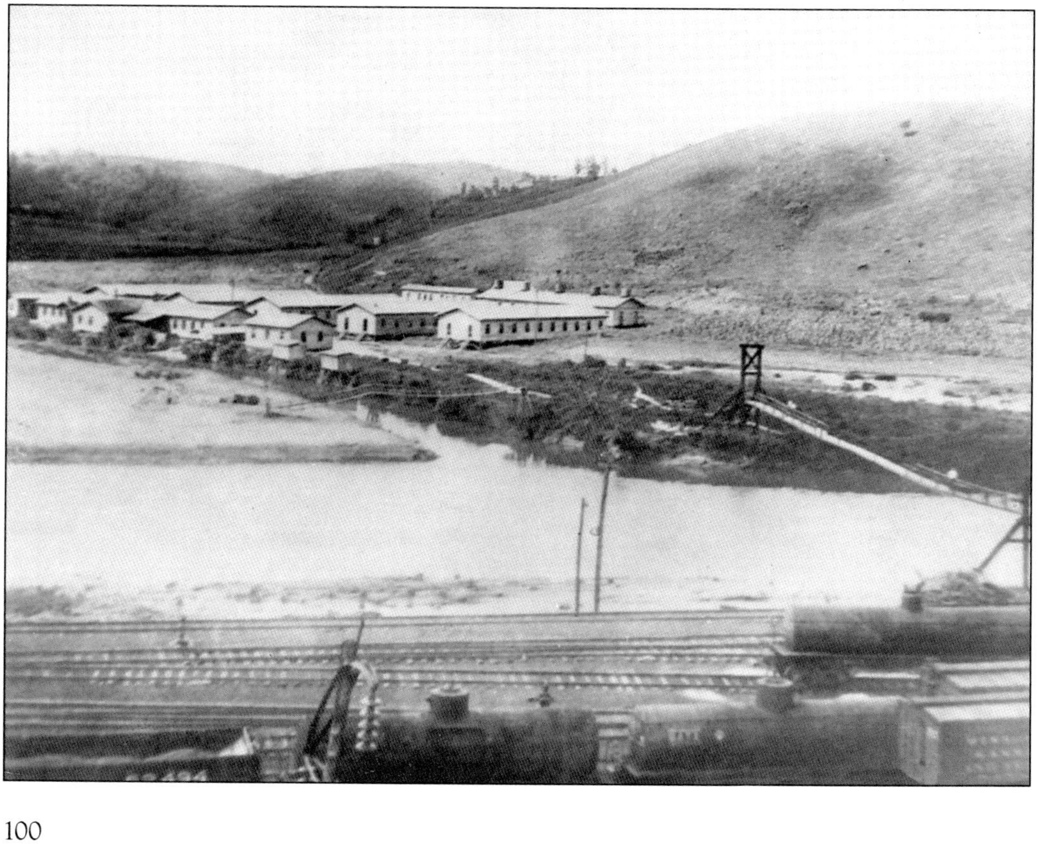

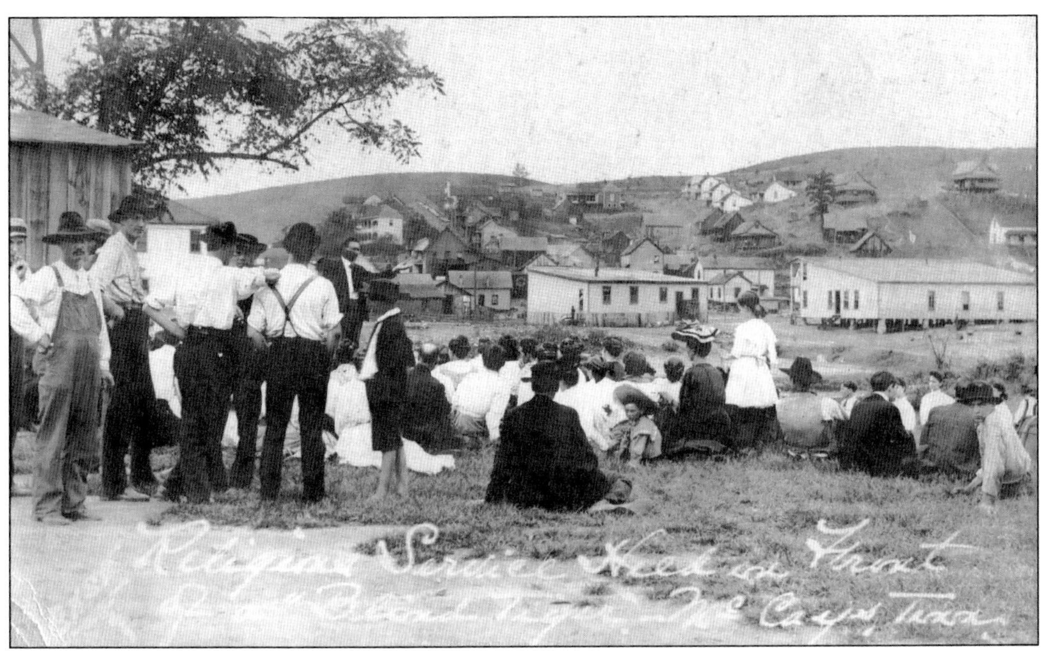

According to oral tradition, the site of the "blind tiger" (the local source for illegal alcoholic beverages) at McCays is thought to have been at one end of the labor-camp footbridge. The c. 1900 photograph above of a religious service at the blind tiger, however, seems to have been taken near the site of the ferry landing and footbridge on the McCays side of the river. Several of the houses in the photograph can be identified in the 1904 E.H. Westlake photograph of McCays on page 54, and what appears at first glance to be a river between the people and the town is actually just a bare patch of land. Below, a different angle of the same service apparently shows the blind tiger itself.

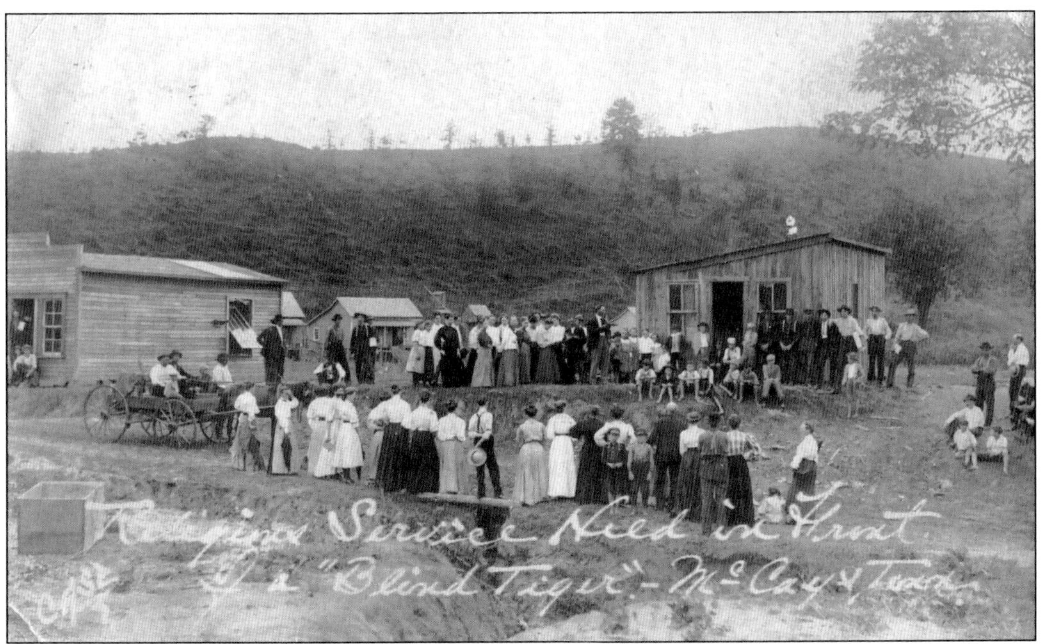

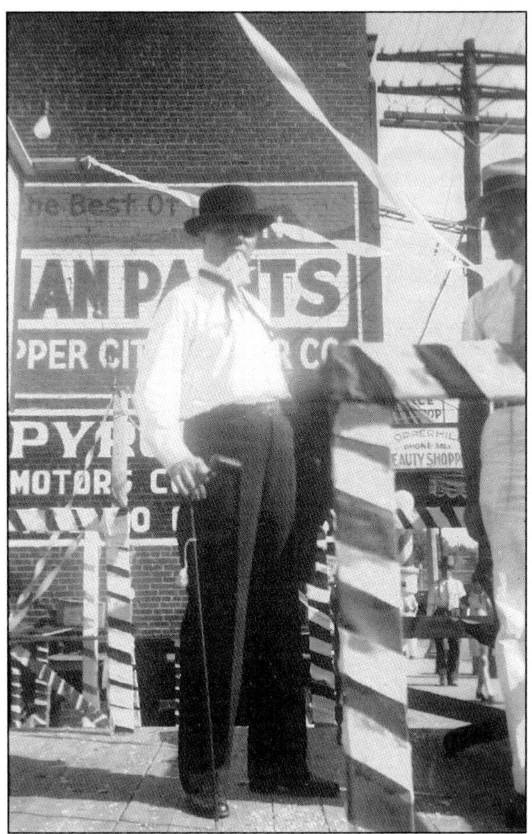

Pharmacist Oscar Tallent, better known to several generations of Copperhill residents as "Doc," opened Tallent Drug Company on Ocoee Street in the 1920s. Doc Tallent was also well known for his sense of humor. In this photograph, he has donned a false beard and dressed as Uncle Sam in celebration of the Fourth of July in Copperhill. (Courtesy of the Howard family.)

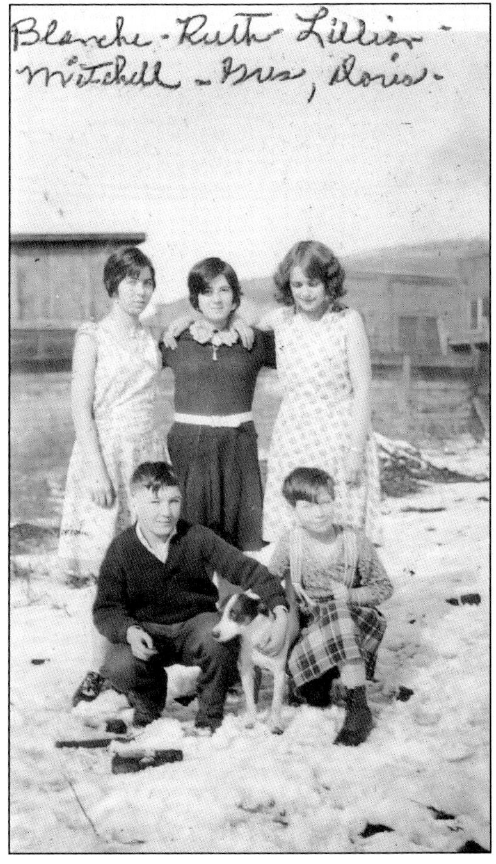

This photograph of Oscar Tallent's children and a friend was probably taken in Copperhill in the 1920s. Kneeling with the dog are Gus (left) and Doris. Standing, from left to right, are Blanche (later the wife of Ralph Howard), Ruth (later the wife of Grady Jones), and Lillian Mitchell. W.G. "Gus" Tallent worked with his father for many years as a pharmacist at Tallent Drug Company. (Courtesy of the Howard family.)

In 1944, at the request of PTA program leaders H.H. "Happy" Hill and his wife, Bess, a tongue-in-cheek style show was held at Copperhill High School. This photograph of Virginia (McCay) Middleton (left) and Polly Morgan, taken by Cliff Parris, was used on posters to advertise the show. On the table is a photograph of Capt. Edward A. Middleton, Virginia Middleton's husband, who was serving overseas with the US Army. (Author's collection.)

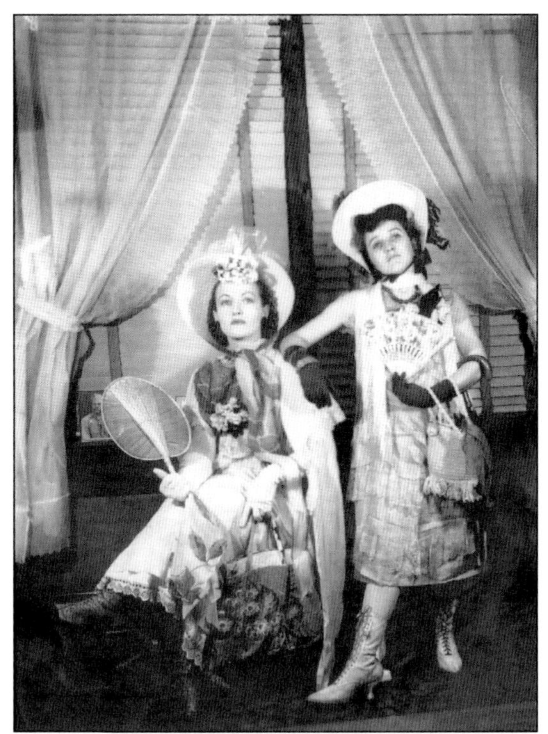

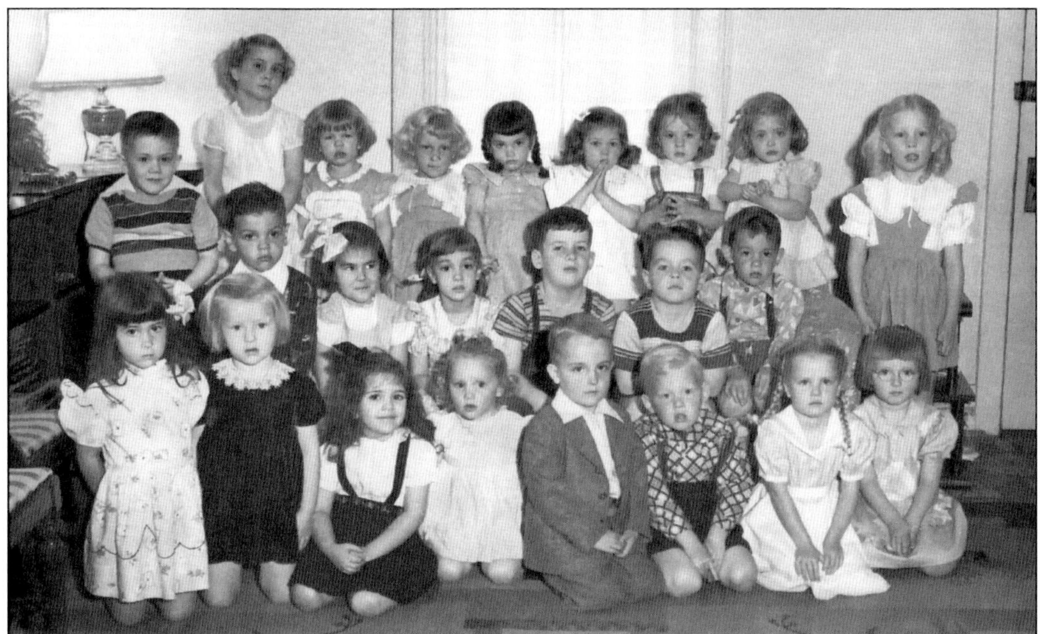

Attending John Roach's c. 1949 birthday party are, from left to right, (first row) Celie Burns, Mary Ann Jones, Betsy Simmons, Alice Ann Seaman, John Roach, Allen Simmons, Ann Branch, and Martha Moncrief; (second row) Bill Hill, Bobby Rook, Judy Keffer, Sally Higdon, James Ward, Jim Moncrief, and Stanley Simmons; (third row) Gail Duke, unidentified, Patsy Rankin, Sally Simmons, Alaire Bretz, Lorraine Akin, Betsy Hill, and Ginny Rankin. (Courtesy of Lorraine Akin Buhrmann.)

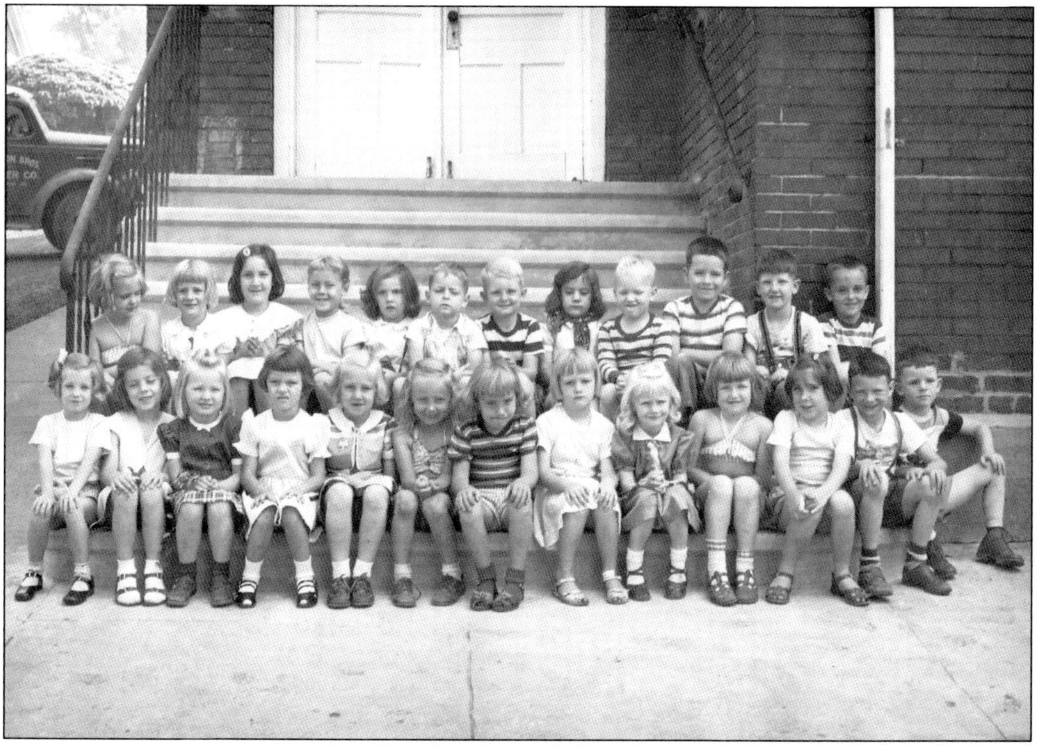

For many years, the First Methodist Church of Copperhill was home to the Basin's only kindergarten. From left to right in this c. 1950 photograph are (first row) Mary Lohr, Lorraine Akin, Marcie King, Mary Helen Howard, Patsy Rankin, Glenna Boye, Sheppie Williamson, Claretta Arp, Virginia Lu Sharp, Martha Moncrief, Susan Spargo, Dale Hamby, and Robert Ward; (second row) Betsy Hill, Leslie McLaine, Linda Stepp, Susan Bailey, Jane Weeks, Charles Swanson, Bill Townsend, Lizzie Estabrooks, Richard Kingman, Jim Moncrief, Quinton Jones, and John Roach. (Courtesy of Lorraine Akin Buhrmann.)

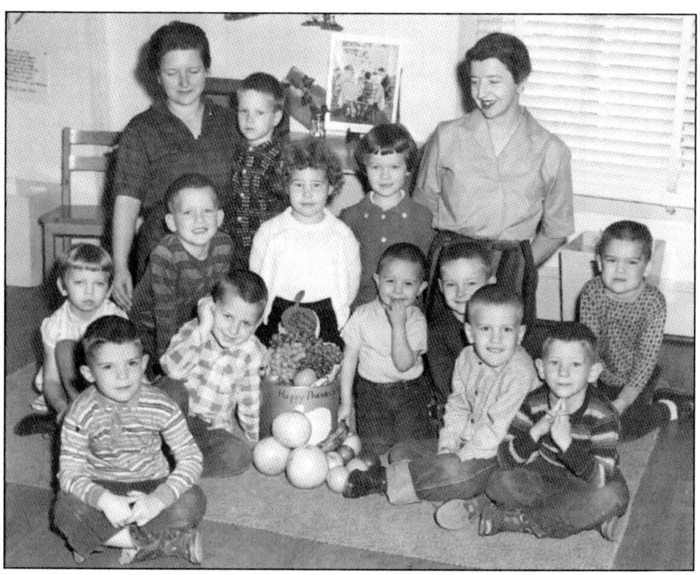

In 1959, the kindergarten children brought fruit to make a basket for a Thanksgiving visit with a homebound local woman, Mrs. I.M. Graham. The children are, from left to right, (first row) Charles DeWitt, Charles Walden, Randy Loudermilk, Bob Lynch, Gene Middleton, Dan Lohr, and Charles Layne; (second row) Amy Quintrell, Ricky Marlette, Rene Bidez, Kay Ballew, and Joetta McCarter. The teachers are Delores Walden (left) and Dorrace Lohr.

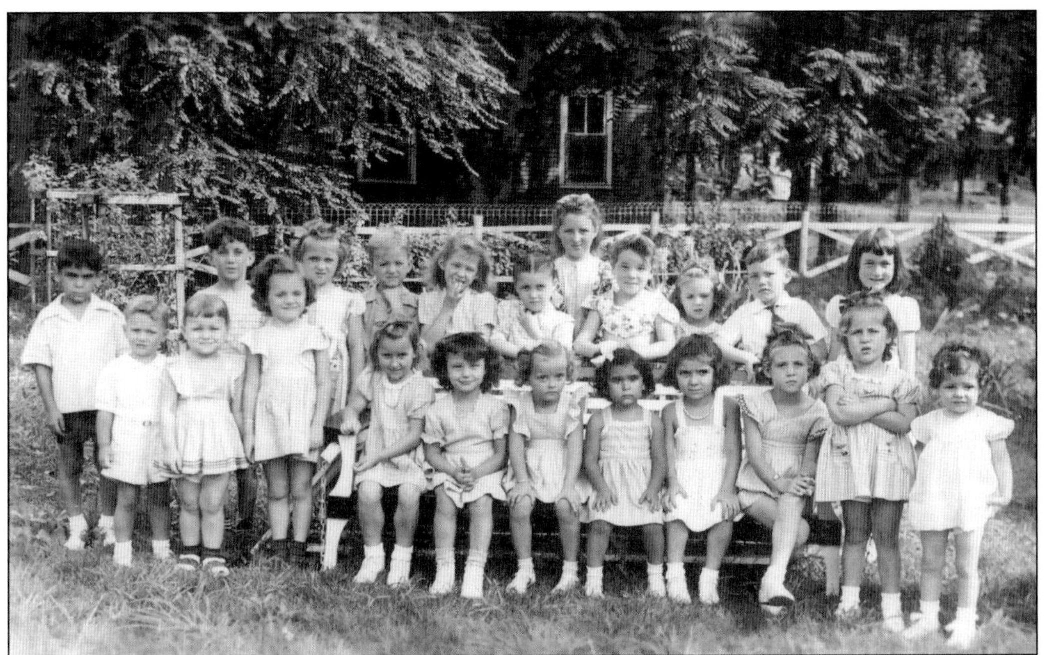

When Peggy Walter celebrated her birthday in 1943, the party hosts somehow managed to wrangle 22 small, cake-fueled children together for the photograph above. From left to right in their party finery are (first row) George Wayne Shibley, Gene Carroll, unidentified, Peggy Walter, Rebecca Adams, Louise Hyde, Carolyn Kaylor, Sonya Shibley, Theresa Jabaley, Judy Spencer, Carolyn Jones, and Sharon Carroll; (second row) Ferris Maloof, Peggy Jones, unidentified, Carleton McCay, June Ann McCarter, Maudine Patterson, Mary Louise Howell, Sandra Rippy, George Cobb Jones, and Mary Nan Higdon. A few years later, Peggy (right) and her Copperhill ballpark neighbor Amelia Dalton pose for the fanciful photograph at right against a painted backdrop. (Both, courtesy of Peggy Walter Kilpatrick.)

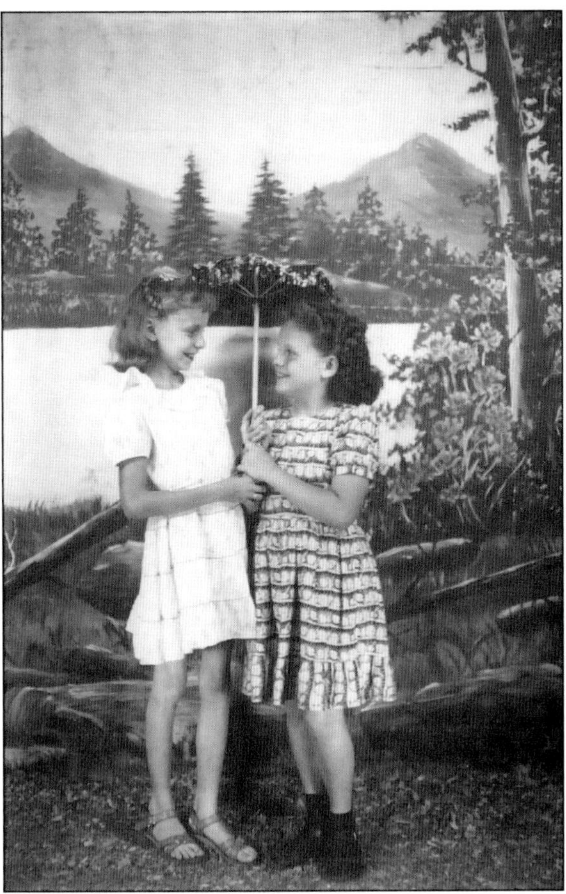

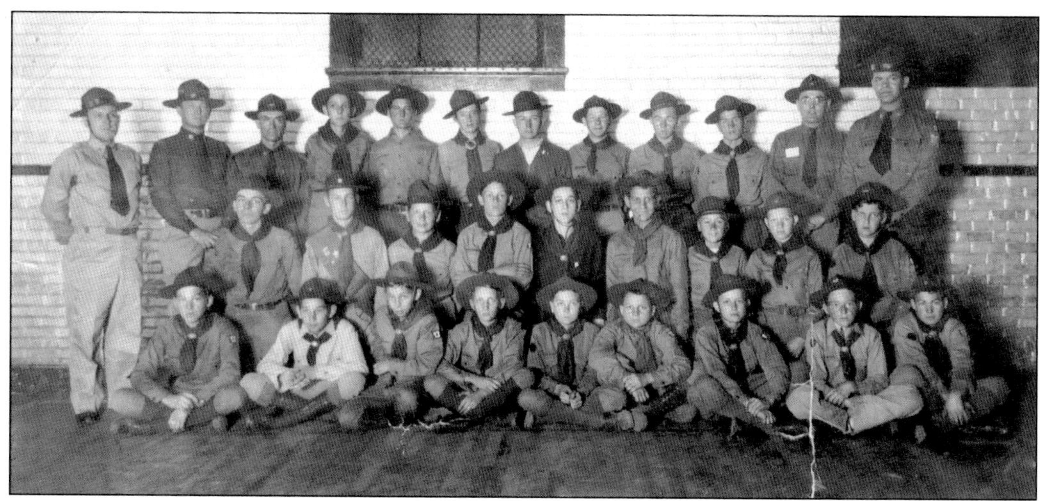

Copperhill's Boy Scout Troop 13 poses for a formal photograph in the YMCA gym in 1941. Assistant scoutmaster Carl Stepp is third from left in the third row, Roy Hughes is sixth, and R.E. "Bobby" Barclay Jr. is seventh. In the second row, Bill Dickey is fifth from left, Eddie Quinn is seventh, and Bob Stanley is eighth. The sixth boy from left in the first row is Paul Stepp. (Courtesy of R.E. Barclay Jr.)

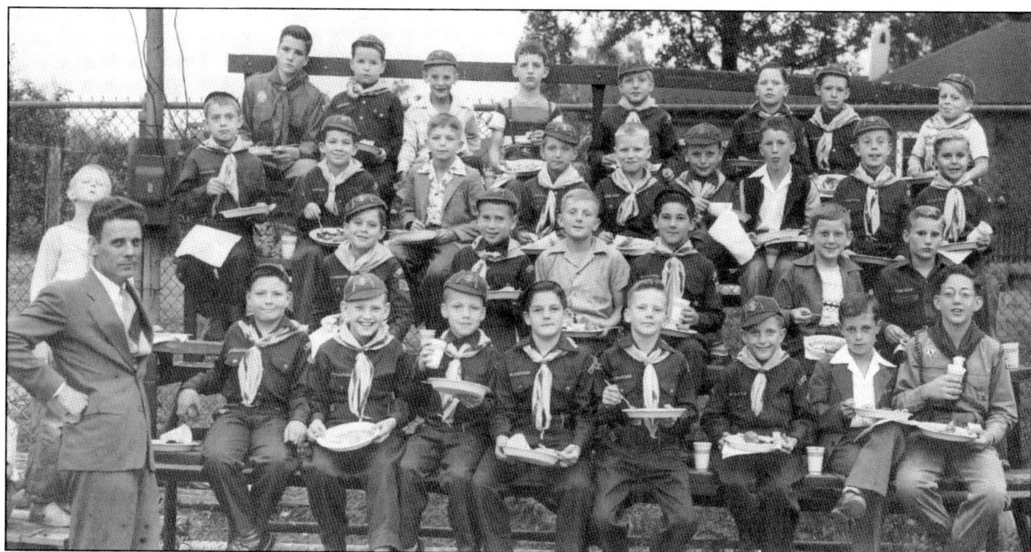

In the early 1950s, scouting in the Basin was still going strong. This Cub Scout troop included boys from both Tennessee and Georgia. From left to right are (first row) scoutmaster Guinn Craig, unidentified, Billy Henry, Roy Loudermilk, Jerry Stepp, Meredith Ryder, Doug Foster, Jimmy Woodall, and Ferris Maloof; (second row) Joe Willis (standing), Ronnie Clay Miller, Steve Carver, unidentified, Joe Stepp, Billy Brackett, and Donald Panter; (third row) Roger Ash, David Abbott, Doug Wilson, Harmon Smith, Michael Derreberry, Billy Dalton, Jimmy Louis Panter, Joe Ray Fortner, and Ronald Ratcliff; (fourth row) Bob Ritchie, Jim Moncrief, unidentified, James Ward, Gene Carroll, Jon Earl Mason, Leland Thomas, and Gilbert Ratcliff. (Courtesy of Laura Ruth Wilson.)

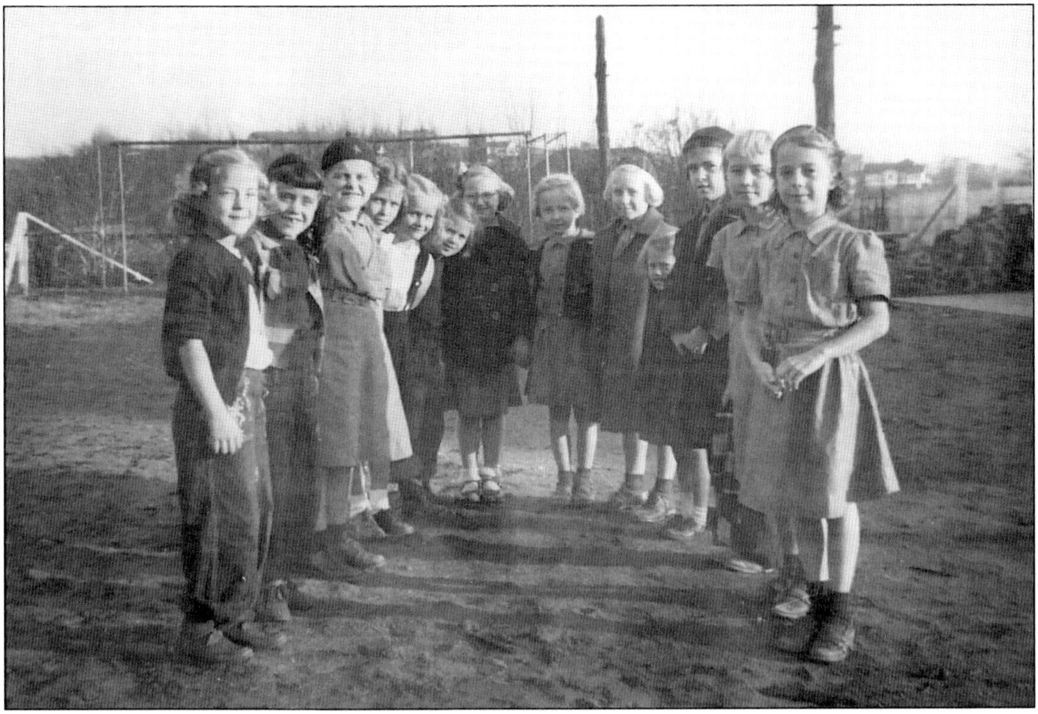

Scouting in the Basin was equally popular with girls. In this mid-1950s photograph of the Copperhill Brownie Scout troop are, from left to right, Susan Bailey, Brenda Jean Gilliam, Marilyn Keenum, Loretta Carter, Claretta Arp, Alaire Bretz, Martha Moncrief, Glenna Ruth Boye, Patsy Rankin, Patsy Bretz, Celie Burns, Sheppie Williamson, and Lorraine Akin. (Courtesy of Lorraine Akin Buhrmann.)

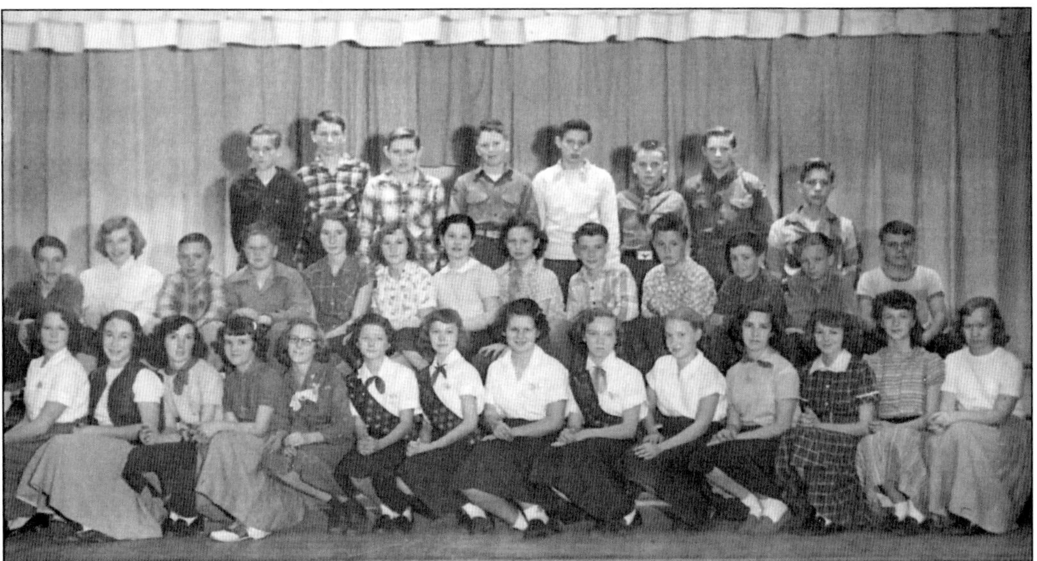

This photograph of the 1953–1954 seventh-grade class at Copperhill School was apparently taken on the same day as a Girl Scout meeting. In uniform at the center of the first row, from left to right, are Harriet Frye (fifth from left), Glenda Sue Panter, Shirley Collis, Jeannie Quinn, Alice McGill, and Janice Duncan. (Author's collection.)

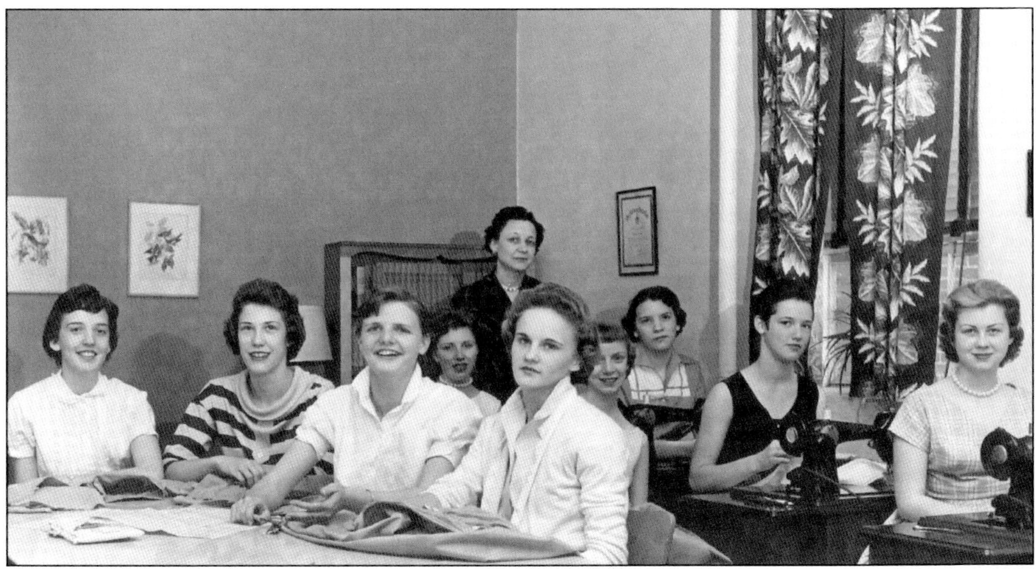

In 1956, members of the home economics class at CHS formed a nurses' aides group and bought materials to sew their own uniforms. Seated from left to right in this photograph are Patricia Neal, Carleton McCay, Helen Townsend, Lynne Ellen Derreberry, Sandra Panter, Amelia Dalton, Mary Louise Howell, Janelle Kimsey, and Carolyn Clyburn. In the background is home economics teacher Marguerite German. (Courtesy of Peggy Walter Kilpatrick.)

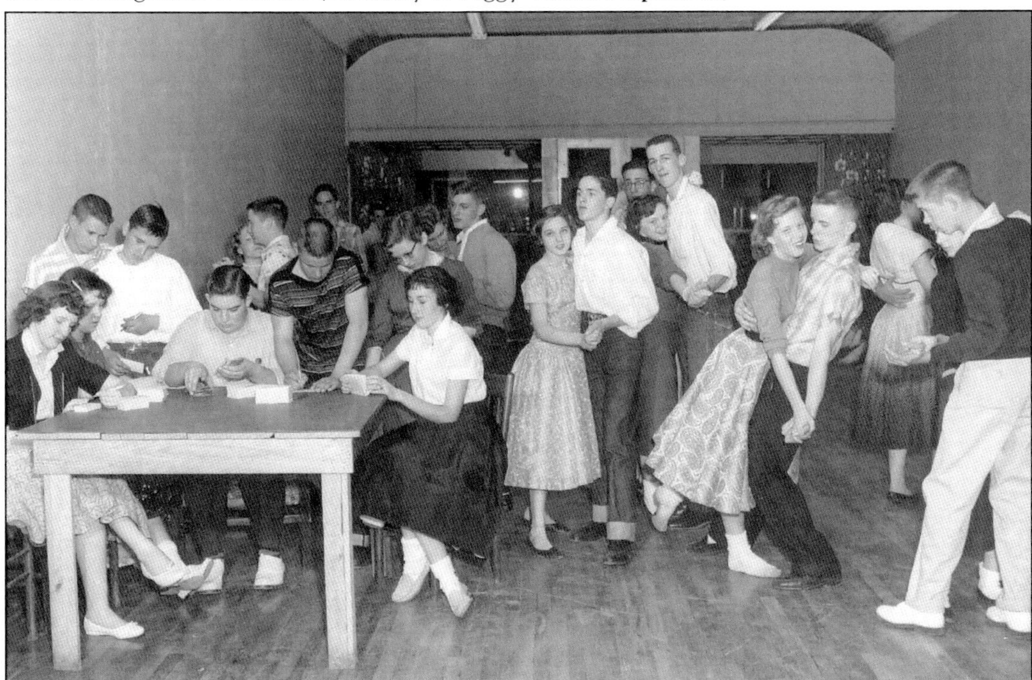

The first Teen Canteen was opened in a downtown Copperhill building in 1956. Among the members seen here are Sue Coffey, Ruth Ann Coffey, Homer Johnson, Mary Nan Higdon, Steve Carver, Albert Carras, Richard Wagner, Barbara Ann Self, Jeannie Queen, Tommy Stanley, Joe Russell, Roger Ash, Sandra Nance, Patricia Godfrey, Ferris Maloof, Peggy Walter, Barry Akin, Annette Howard, Billy Dalton, Patricia Roach, and Russell Amburn. (Courtesy of Bill Dalton.)

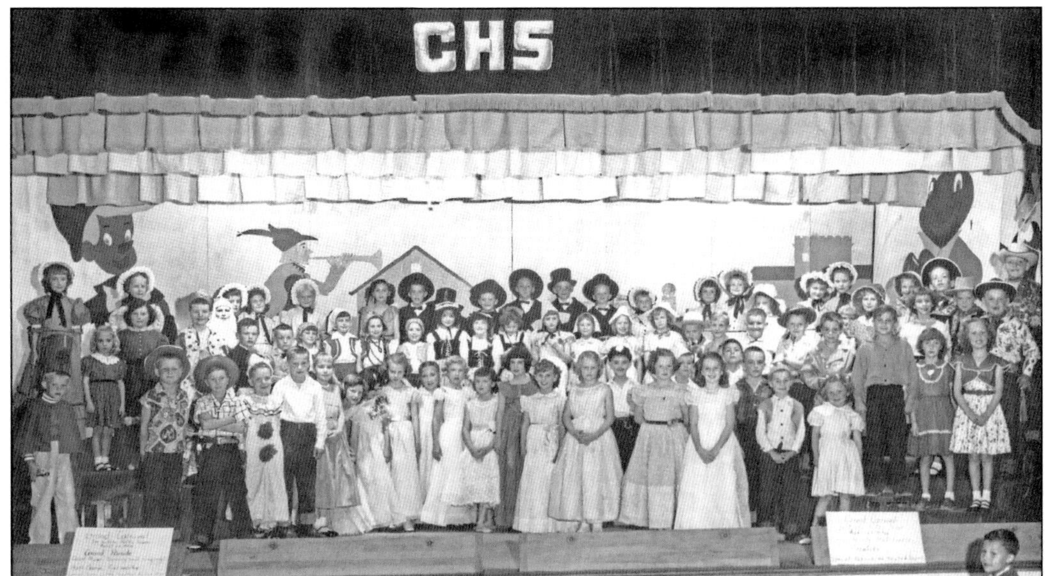

The grammar school operetta at Copperhill School in April 1954 advertised itself as *Grand Carnival in Little Folks Town*, complete with a grand parade, music, dancing, singing, a ball game, and fireworks. Operetta casts were typically composed of children from two or three different grades. (Courtesy of Lorraine Akin Buhrmann.)

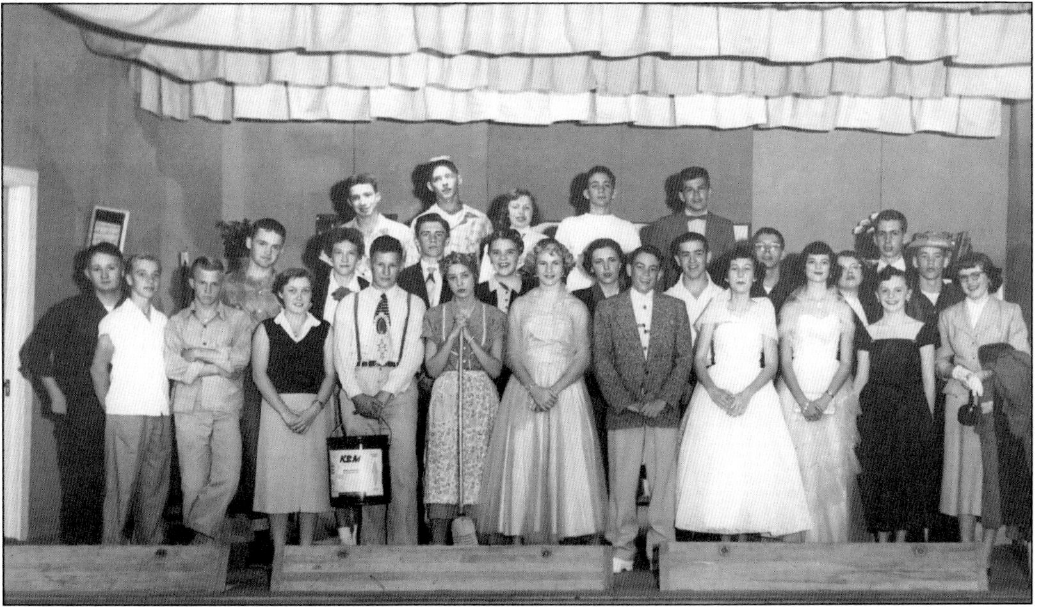

From left to right, the cast of the 1955 sophomore class play at Copperhill High School are (first row) Gordon Luther McMahan, Marvin Lee Marchman, James "Bee" Hicks, June Ann McCarter, James Howard Crawford, Angela Nazerian, ArSula Thomas, Tony Middleton, Mary Nan Higdon, Jane Ballew, Peggy Walter, and Barbara Ann Self; (second row) Wally Arp, Carleton McCay, Edwin Light, Mary Louise Howell, Patricia Roach, Marvin Ferguson, James Roy Mitchell, unidentified, Britt Burns, and George Cobb Jones; (third row) Tom Beavers, Glen Towery, Lynne Ellen Derreberry, Ferris Maloof, and Mike Morgan. (Courtesy of Peggy Walter Kilpatrick.)

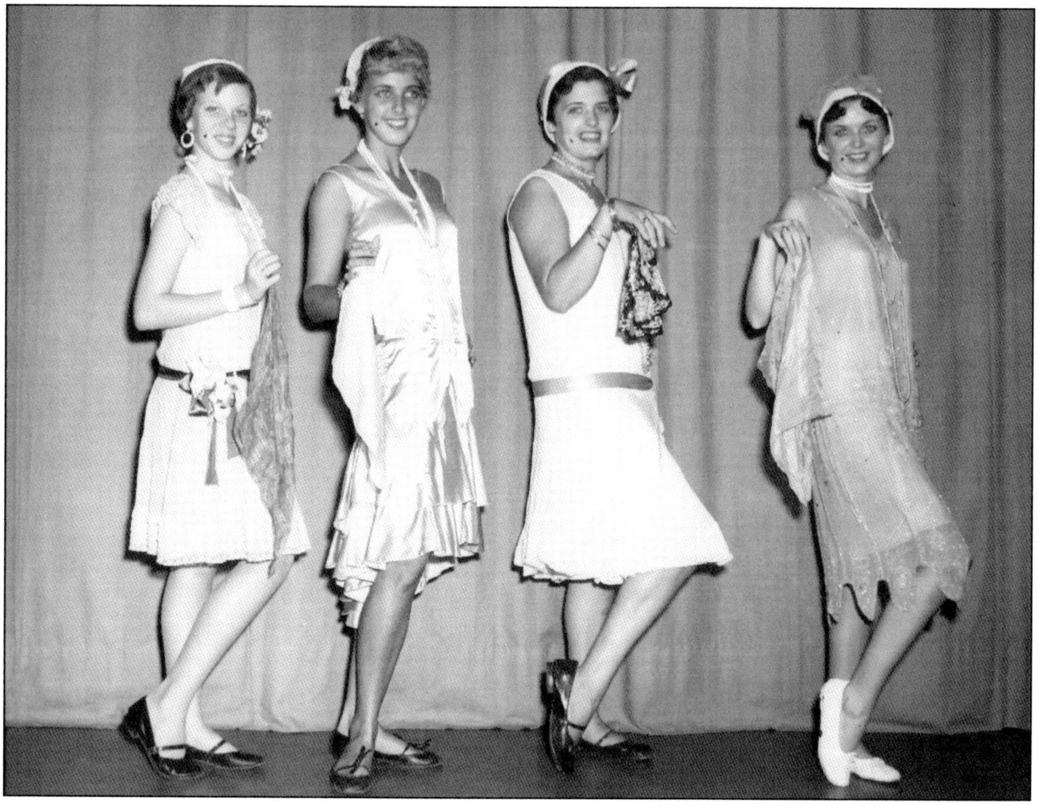

In 1953, and again in 1954, the communities of the Basin held weeklong Second Century celebrations to commemorate the opening of the first copper mine in 1850. Dressed as flappers (above) for a drama at Ducktown School in 1953 are, from left to right, Annette Howard, Eve Heriot, Harriett Hill, and June Dalton. Below, in 1954, a float carrying copper queen Nancy Keller and runner-up Cecelia Jones leaves Five Points in Ducktown at the beginning of the parade route to Copperhill. (Above, courtesy of Bill Dalton; below, courtesy of Alice-Ann Seaman Ferderber.)

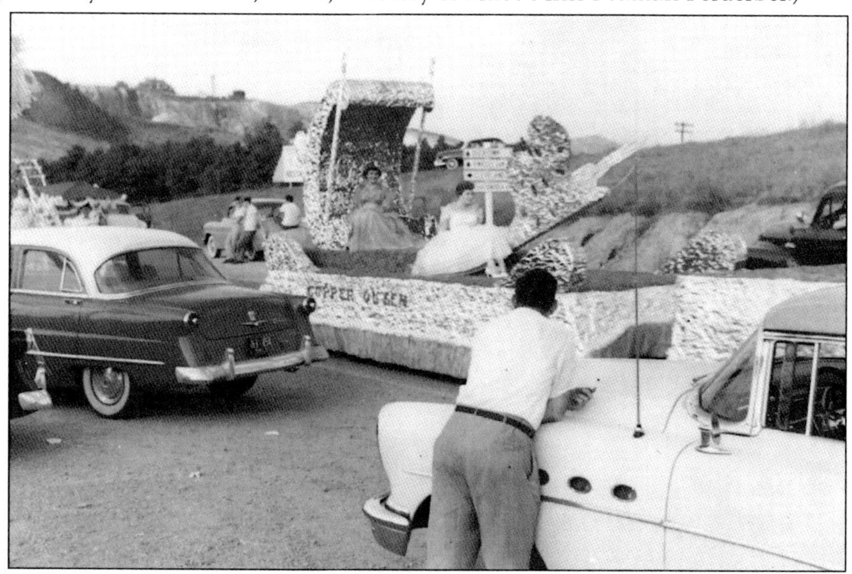

During his political career, Tennessee senator and 1956 vice presidential nominee Estes Kefauver visited the Basin on several occasions. In this photograph, he poses with Tommy Walker (left) and Joe Woodward, who is apparently wearing the senator's trademark coonskin cap. In the background is Donnie Wilson. The photograph may have been taken during the 1954 Second Century celebration, at which Senator Kefauver was a featured speaker. (Courtesy of Peggy Walter Kilpatrick.)

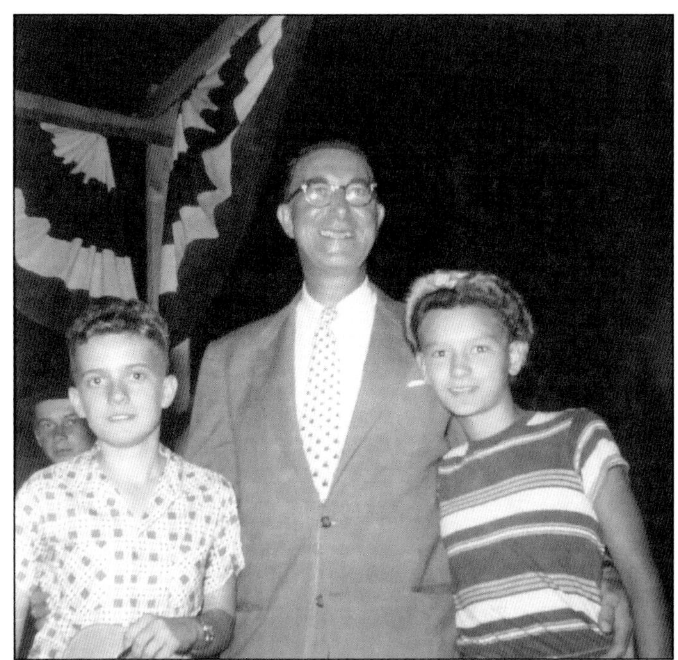

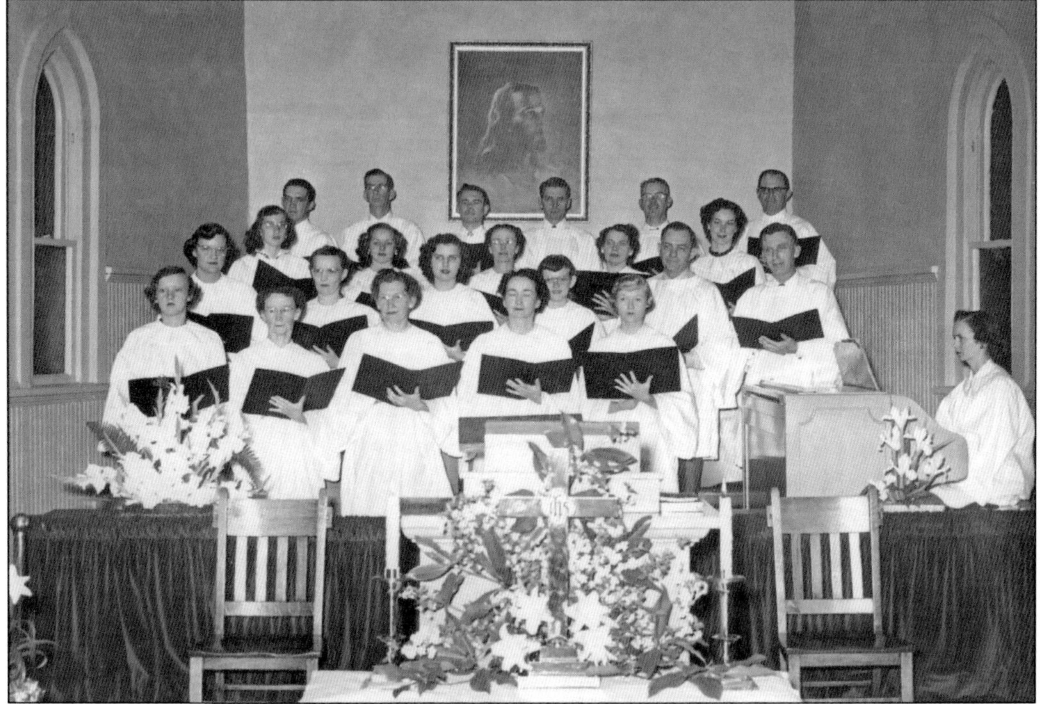

Choir members at the First Methodist Church of Copperhill in the early 1950s are, from left to right, (first row) Elizabeth Rankin, Imah Shelton, Anna Panter, Eunice Taylor, and Mary Amburn; (second row) Rethel Pack, Virginia Middleton, Martha Townsend, Barbara Self, Ben Wynn, and C.F. "Spot" Sharp; (third row) Nellie Posey, Loretta Loudermilk, Gertie Frye, Esther Sharp, and Margaret Jean Taylor; (fourth row) Tommy Posey, Al Sivils, Carl Panter Jr., James Goodman, J.S. Akin, and W.C. Posey. Sara Williamson is at the organ. (Author's collection.)

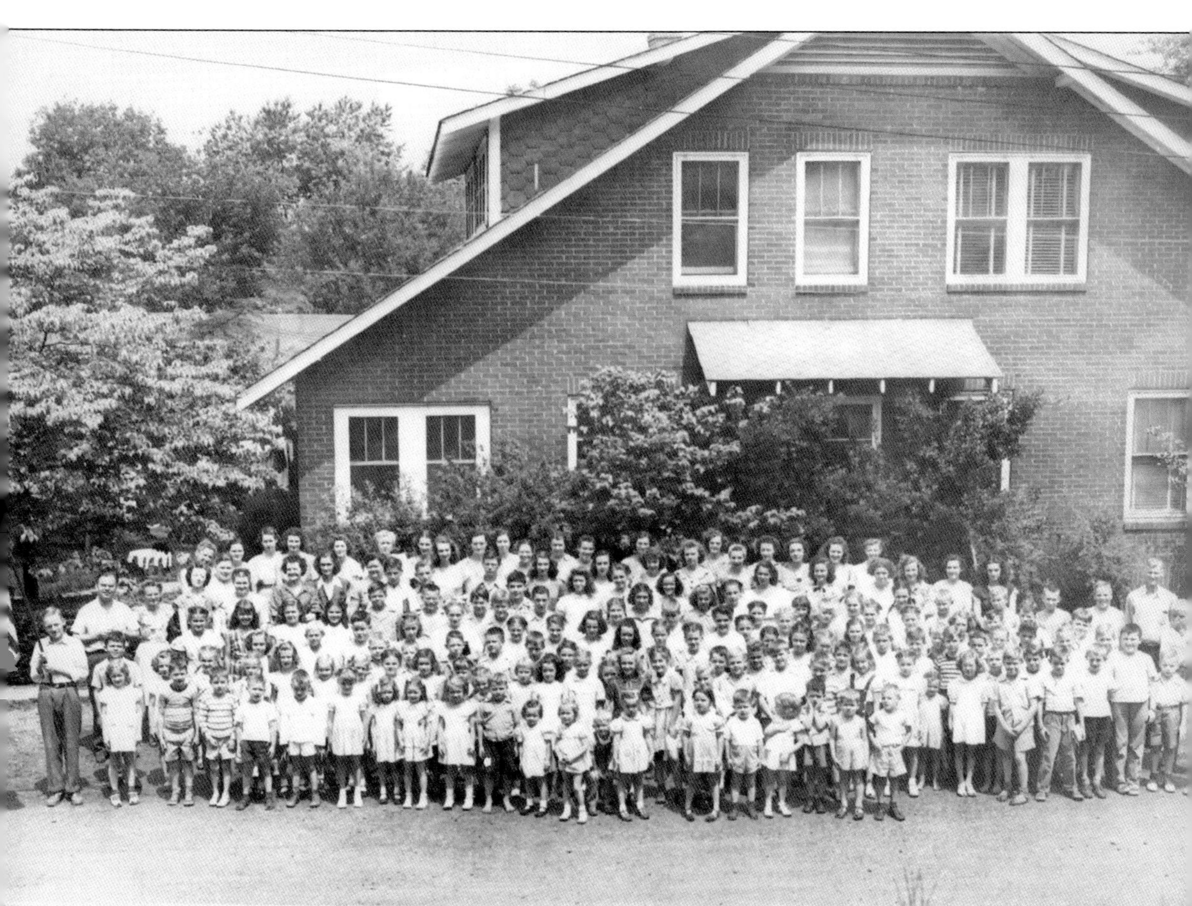

This late 1940s photograph of a Vacation Bible School group at the First Baptist Church of McCaysville was taken in the yard beside the church parsonage. Some of the children who can be identified are Lamar Sisson (holding the flag), David Abbott, Judy Chapman, Amelia Dalton, Benny Dalton, Billy Dalton, Zellie Guinn Earnest, Haines Hill, Harriett Hill, Mary Louise Howell, Leon Howell Jr., Wallace Kell, Dan Keller, Nancy Keller, Gordon Luther McMahan, Tony Middleton, Mike Morgan, Frank Newman, Patricia Roach, Jackie Townsend, Peggy Walter, Tommy Walker, Gary Williams, Linda Williams, and Nancy Williams. Among the adults are Blanche Howard, Mary Lee Howell, Rachel Higdon, Thelma Middleton, Elizabeth Panter, and Edna Weeks. The city of McCaysville in Georgia adjoins the city of Copperhill in Tennessee and has always been considered a part of the Copper Basin. (Courtesy of Peggy Walter Kilpatrick.)

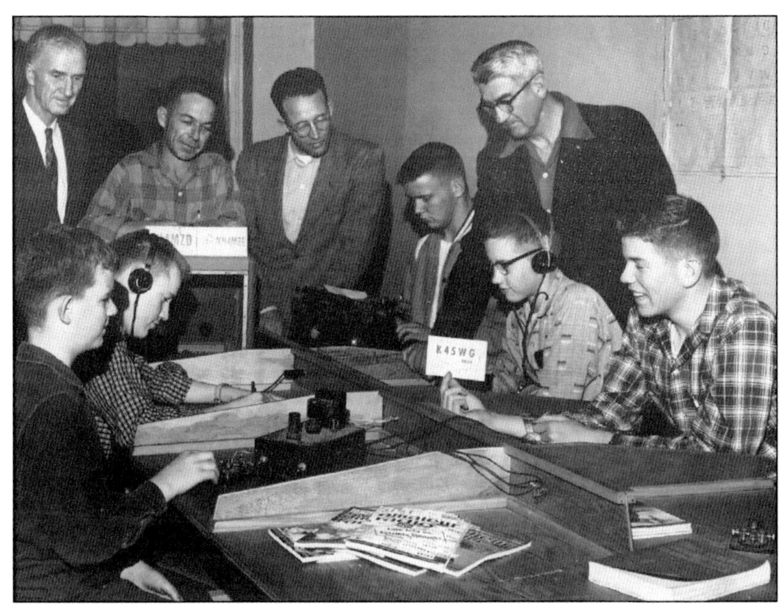

In this 1958 photograph, members of the Copper Basin Amateur Radio and Electronics Club practice code transmission. From left to right are (seated at table) Robert Ward, William Stiles, Richard Wagner, Albert "Rusty" Sivils, and Haney Howell; (standing) Louis Ward, J.M. Griner, Joseph Huskey, and Robert Kilpatrick.

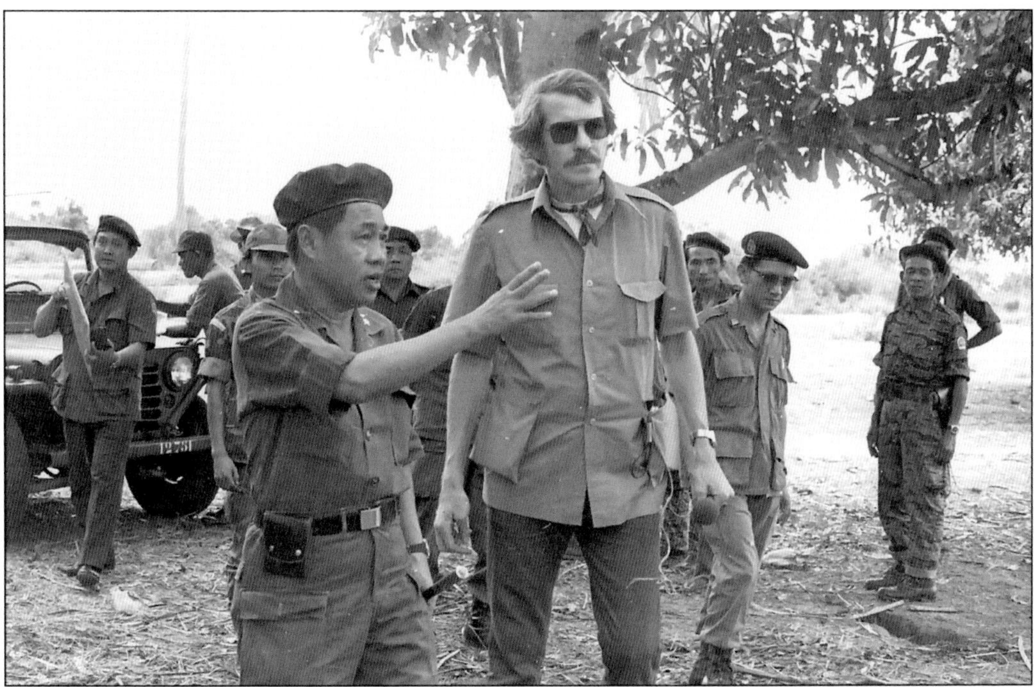

Haney Howell's early enthusiasm for radio led him to a long career in broadcasting, both as on-air talent and as a producer, including two years as a foreign correspondent and Saigon bureau chief for CBS News during the Vietnam War. This photograph was taken when he was reporting from Cambodia near the end of the war. Howell also covered Pres. Gerald Ford's historic visit to China in 1975. In 1988, after stints at ABC Radio in New York and KUSA in Denver, he accepted a teaching position at Winthrop College in Rock Hill, South Carolina, where he created the broadcast journalism major. (Courtesy of Haney Howell.)

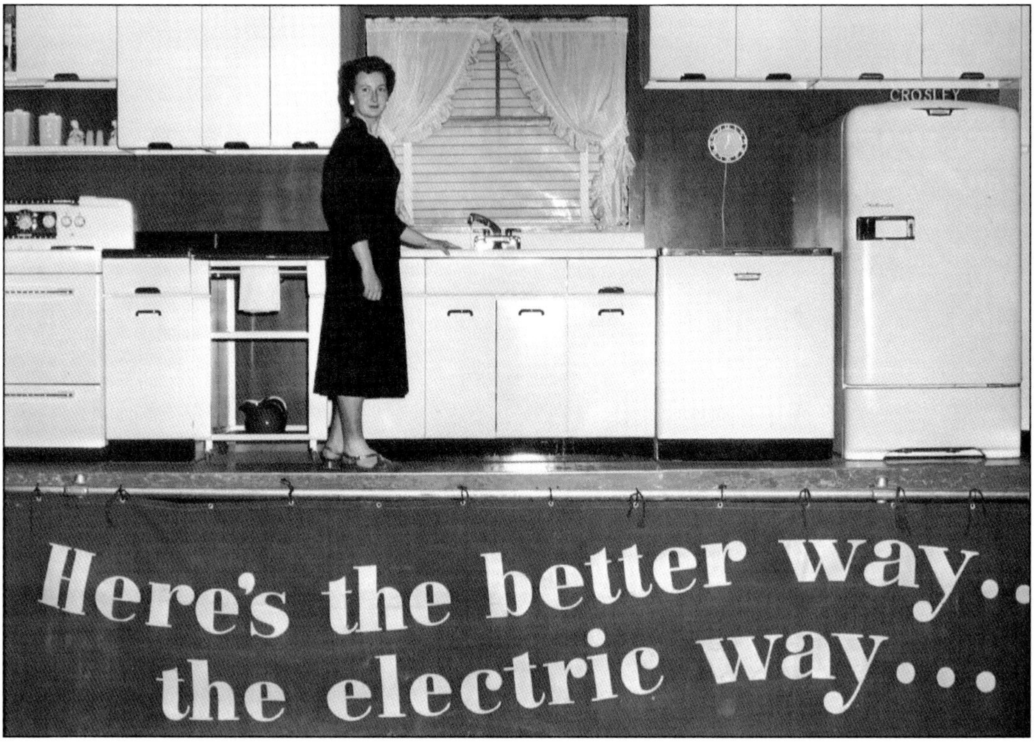

In the 1950s, displays at the annual meetings of Tri-State Electric Membership Corp. (EMC) celebrated the wonders of electricity, which had been available in the major towns for several decades but had arrived only recently in some smaller local communities. Among the duties of employee Margaret Jean Taylor (above) was to help local homemakers learn to adapt their traditional cooking techniques to their new electric appliances. Biscuits, formerly baked in wood stoves, proved to be a problem until Taylor got the idea of turning the biscuit pans upside down and baking the biscuits on top. Below, Carl Ed Abernathy of Abernathy Supply Company shows off his company's wares. (Both, courtesy of Tri-State EMC.)

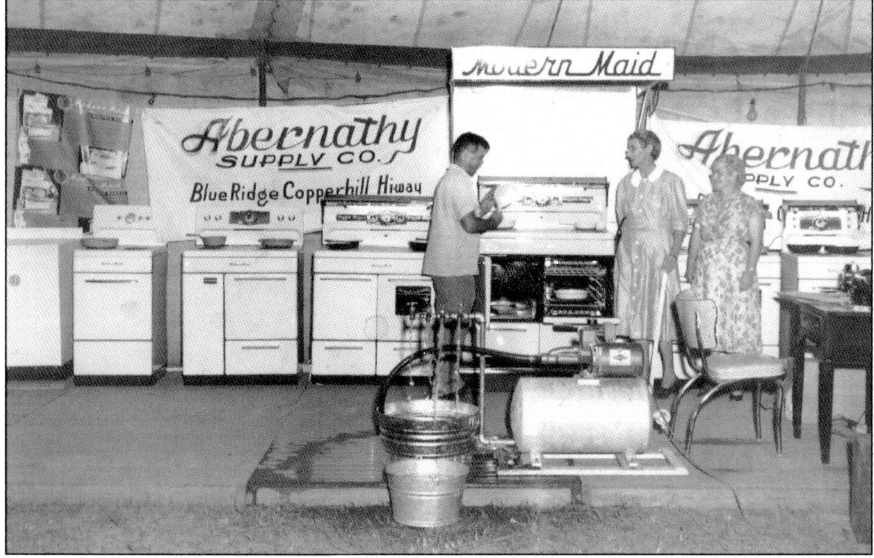

In this 1954 photograph for the employee handbook, Dr. Windom Kimsey consults with "patient" Sue Mitchell at the Tennessee Copper Company hospital. The hospital, opened in 1907, was located next door to the Blue Goose. It remained in operation until the opening of the new Copper Basin General Hospital in August 1955. The photograph was taken by well-known Copperhill photographer Erma Latham.

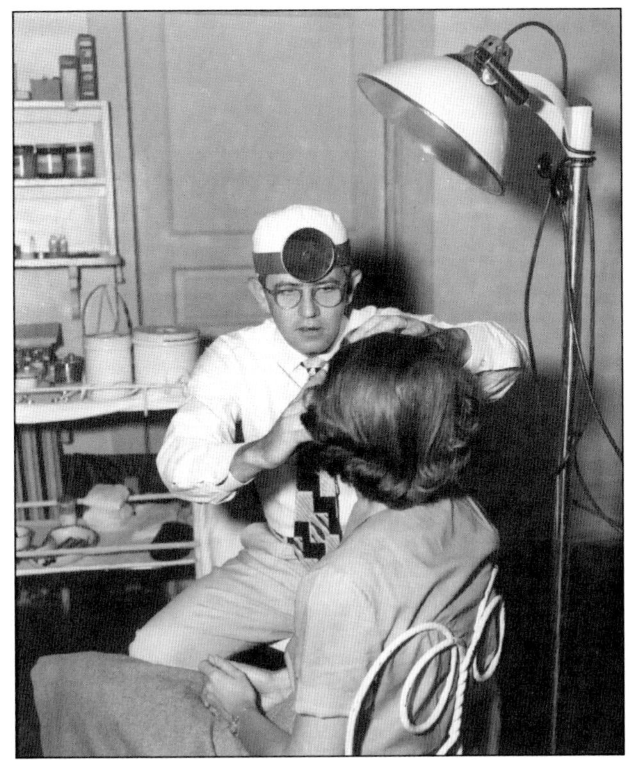

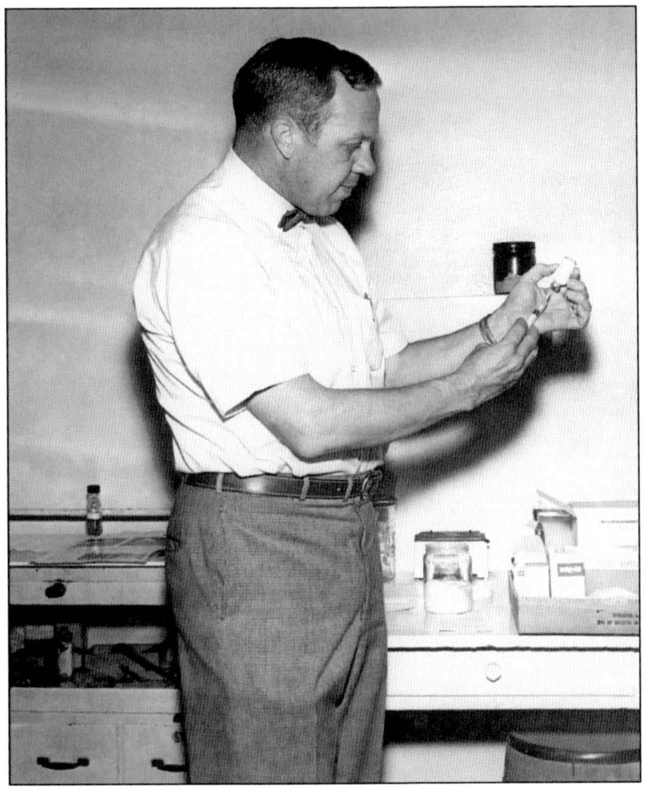

In 1958, Dr. William R. Lee opened his medical practice on Main Street in Ducktown. Until his death in 2015 at the age of 88, he was known as one of the most dedicated and caring doctors in the Basin, as well as one of the most fun-loving. He and his wife, Lorraine, made a total of 64 medical mission trips, 45 of them abroad, and he ran free health clinics in several nearby communities. In the 1970s, he became the founder of Ducktown's popular First Tuesday musical event, in which he participated with enthusiasm. He was also a founder of the Ducktown Basin Museum. In this 1968 photograph, he prepares a syringe for a flu shot.

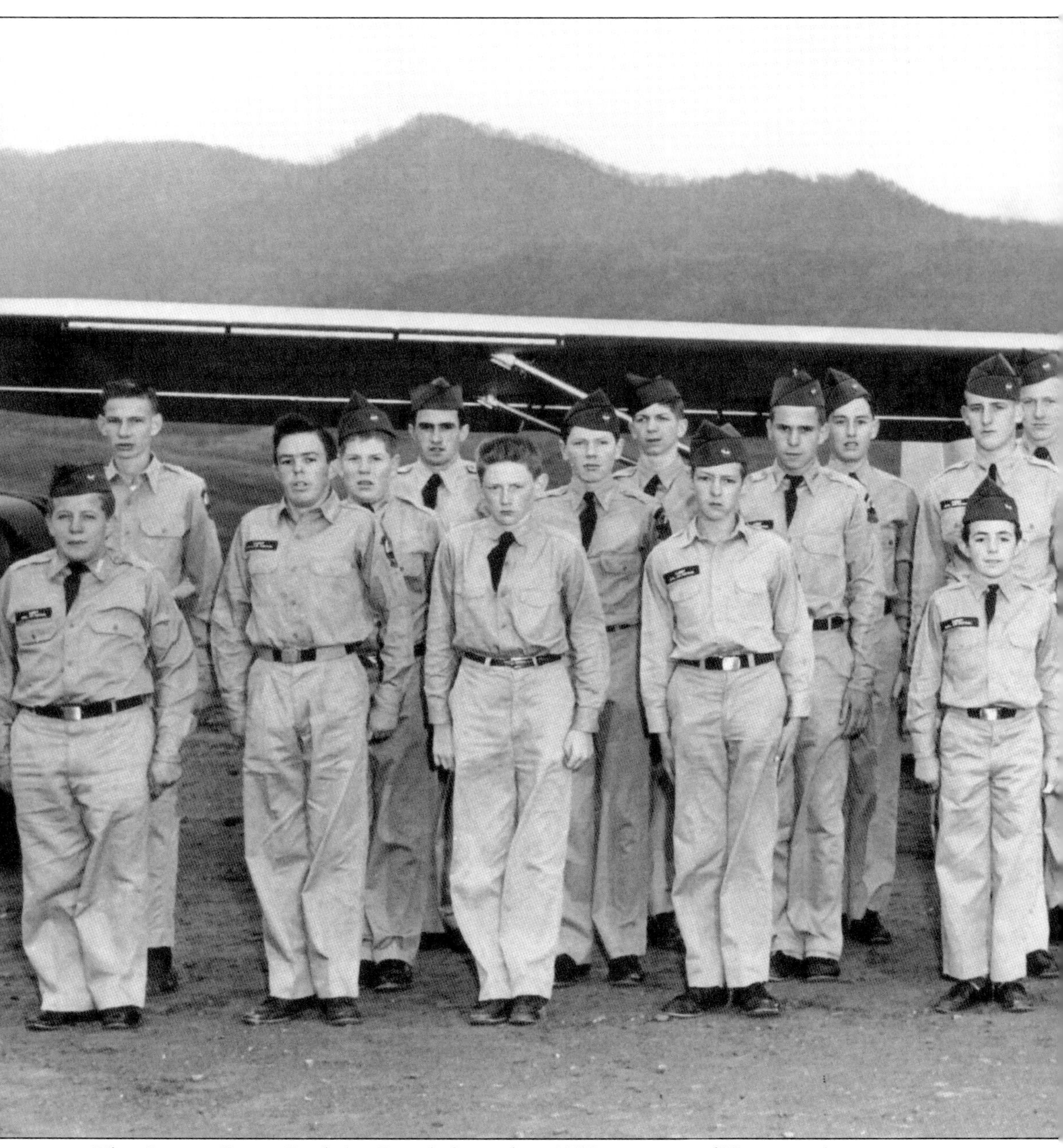

The Copper Basin Civil Air Patrol, formed in 1956, was one of Tennessee's most active units. This L-5 Stinson airplane had been donated by the US Air Force and flown from New Orleans to Copperhill by squadron member Tom Beavers. Cadets pictured on the left, from left to right, are (first row) Gilbert Ratcliff, Maynard "Chubby" Burnette, Freddy Walden, Leland Thomas, and C.L. Burnette; (second row) Joe Davis, Haney Howell, Jon Earl Mason, unidentified, and Bobby

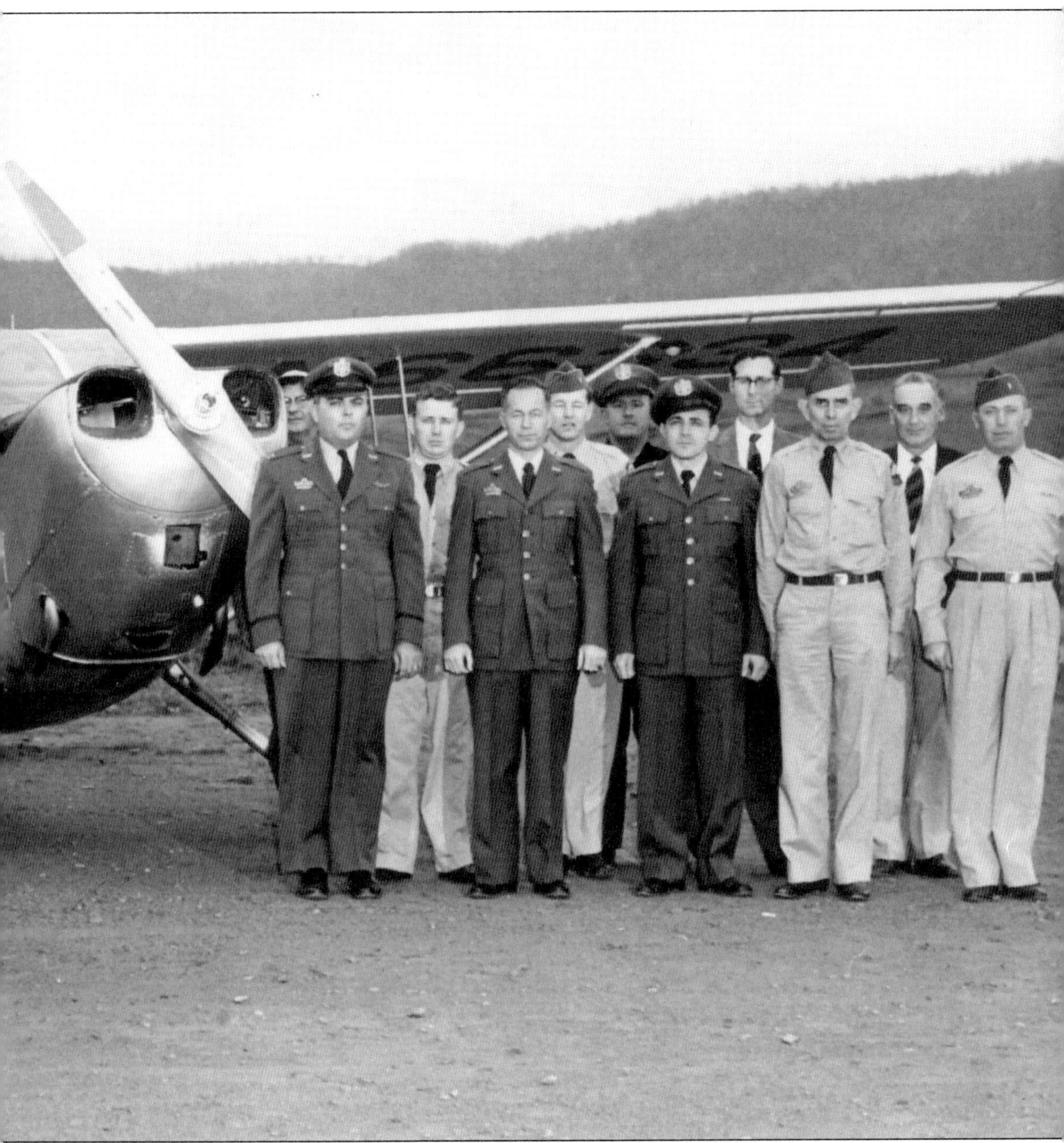

Dickey; (third row) Joe Russell, Douglas Wilson, Jackie Abercrombie, and Marvin Lee Lowery. Adult squadron members pictured on the right, from left to right, are Willard Taylor, captain and commanding officer Dr. J.T. Layne, James R. "Buddy" Finch, unidentified, Lt. Robert Lohr, Lt. Kimsey Campbell, Lt. Emil Greene, unidentified, Glen Weese, George Jones, and Lt. W.L. "Swannie" Swanson. (Courtesy of Howell Layne.)

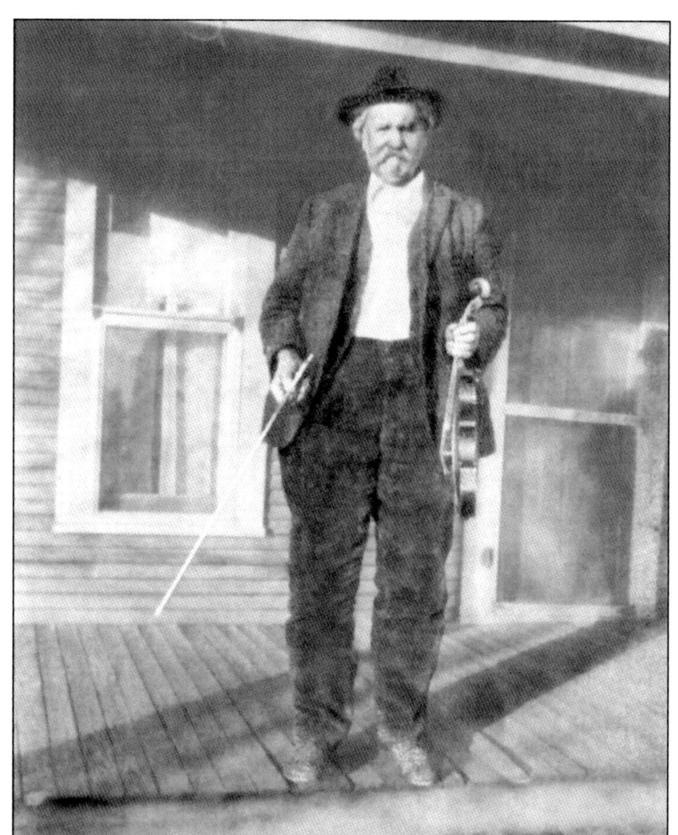

Before the arrival of the railroads, local copper ore was hauled to the railhead at Cleveland, Tennessee, in mule-drawn wagons. The most famous of the early copper haulers was George Barnes, who always took along his fiddle to while away the time on the four-day journey. This photograph was taken in February 1915 on his front porch in Grassy Creek. (Courtesy of the Barnes family.)

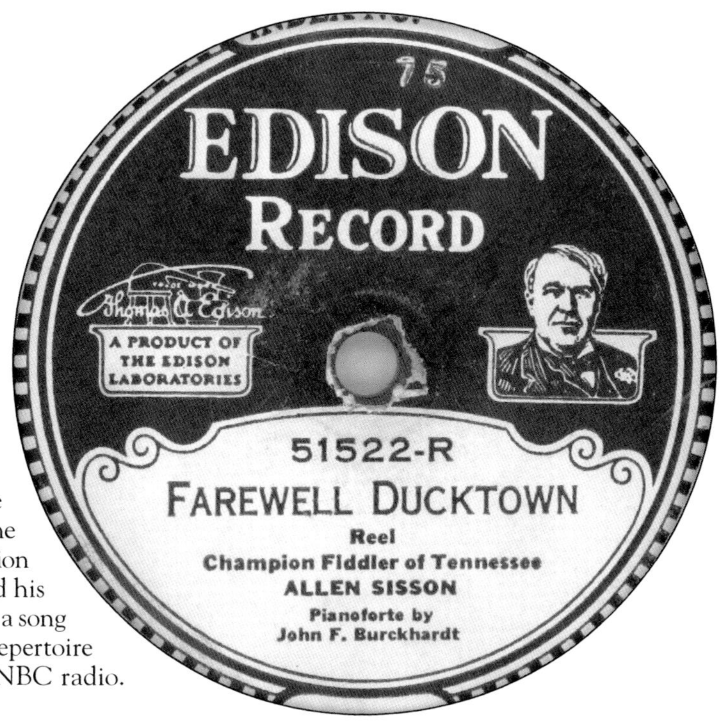

While George Barnes was putting in his final years as a copper hauler, another local fiddle-playing country boy was coming along behind him. Allen Sisson, a section hand foreman on the DSC&I railroad, was an accomplished fiddler by the age of 12 and was named the Tennessee state fiddle champion in 1921. In 1925, he recorded his classic "Farewell Ducktown," a song that became a staple of his repertoire during live broadcasts on NBC radio. (Author's collection.)

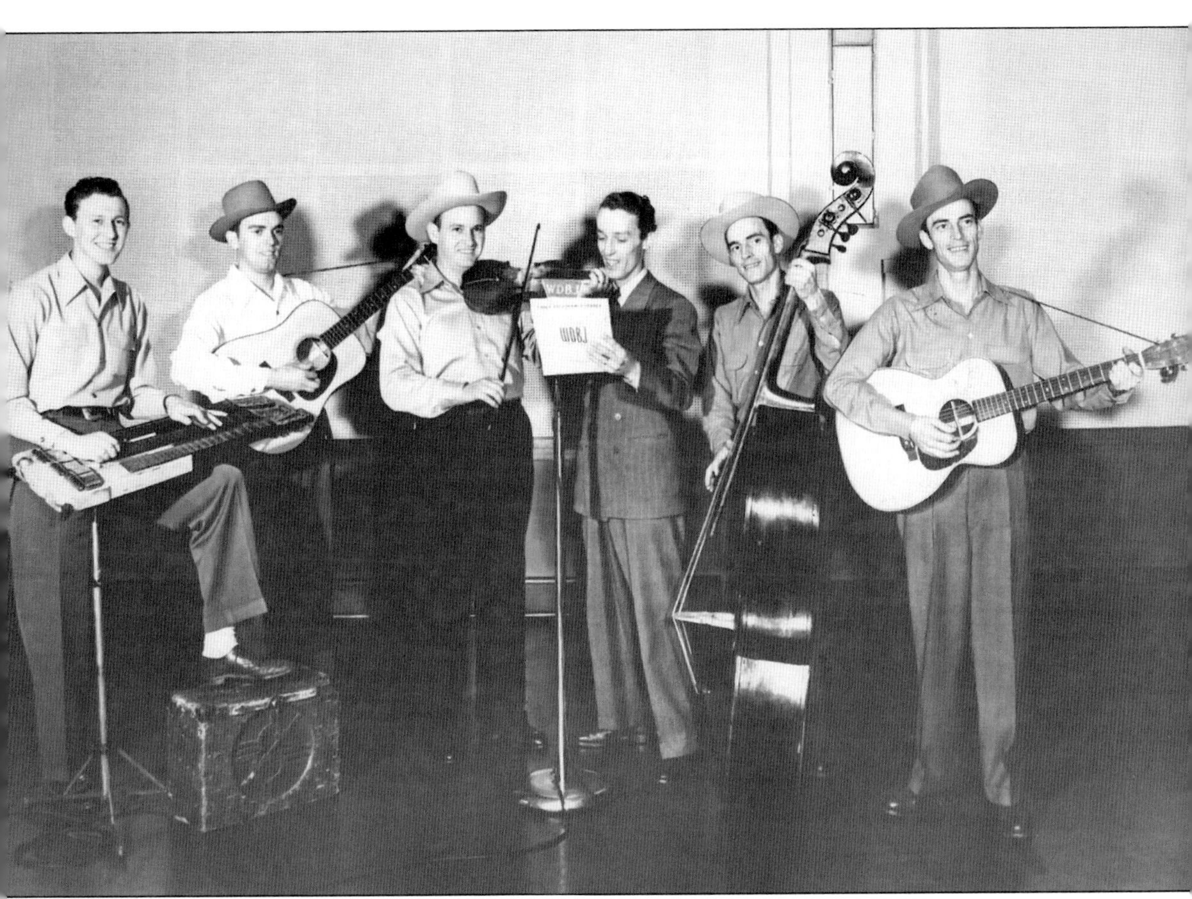

Copper Basin native Tommy Magness (1916–1972), who grew up in a musical family and was performing in public by the age of nine, is widely recognized as one of the definitive bluegrass fiddlers of his generation. Between 1938 and 1952, he played with Roy Hall and his Blue Ridge Entertainers, Bill Monroe and the Blue Grass Boys, and Roy Acuff and his Smoky Mountain Boys, in addition to heading up two bands of his own. He made the first studio recording of "Orange Blossom Special" in 1938, popularized the old mountain fiddle tune that later became "Black Mountain Rag," and recorded what many fiddlers still think is the definitive version of the bluegrass classic "Katy Hill." This photograph of his own band, Tommy Magness and the Orange Blossom Boys, was taken in 1947 at the studios of WDBJ Radio in Roanoke, Virginia. From left to right are Slim Idaho on the three-necked steel guitar, Warren Poindexter on guitar, Tommy Magness on fiddle, WDBJ announcer Dexter Mills, Clayton Hall on bass, and Saford Hall on guitar. (Courtesy of Ralph Berrier.)

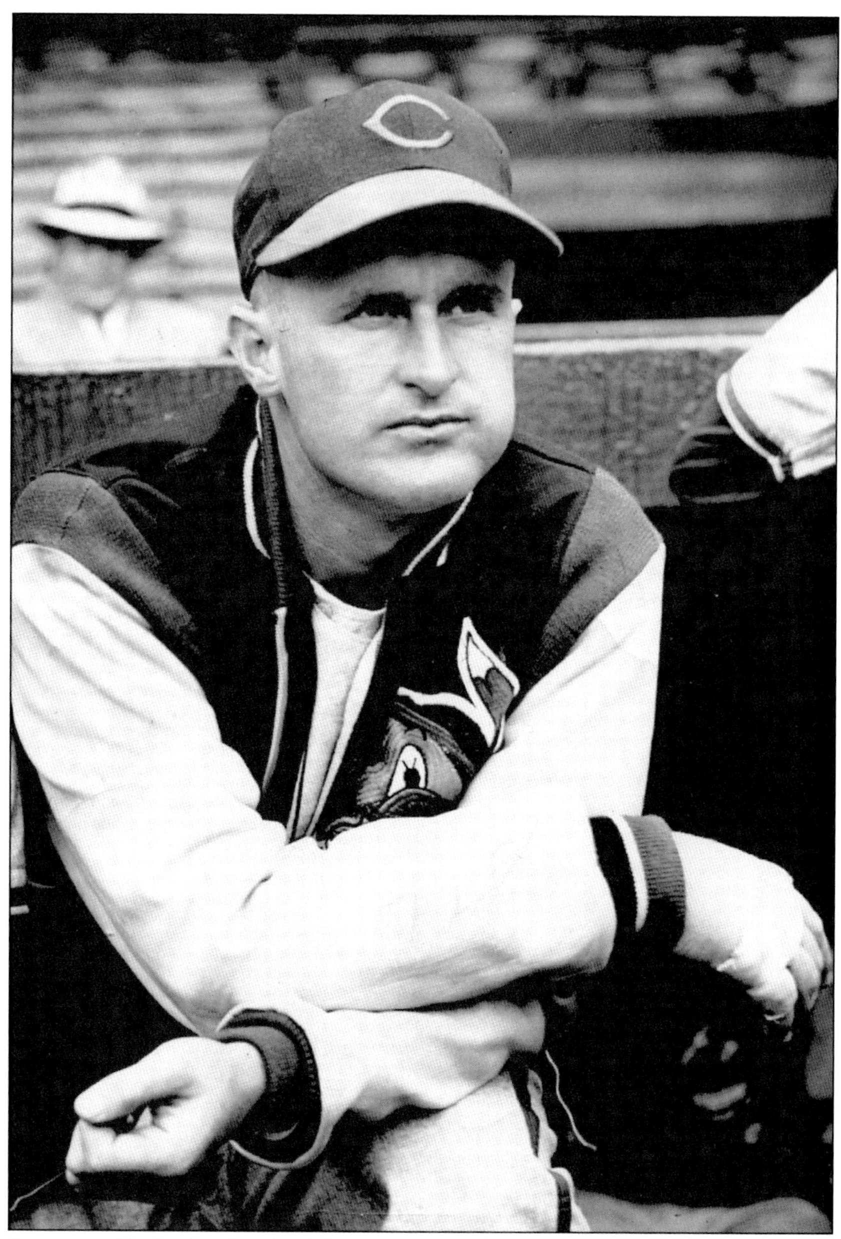

In 1941, after a notable high school athletic career and several years of semipro baseball, 19-year-old Basin native Joe Tipton was signed by the Cleveland Indians. He rose quickly through the ranks of the Indians farm teams and was well on his way to the major leagues when his career was interrupted by a three-year stint in the Navy during World War II. In 1948, he finally found himself in the coveted Indians uniform—and in the World Series, facing the Boston Braves. During the next seven years, he also played with the Chicago White Sox, Philadelphia Athletics, and Washington Senators before retiring from the major leagues in 1954. In addition to his recognized athletic ability, he was famous for his constant and funny game chatter. Apparently, he was particularly good at getting under the skin of serious, focused hitters like Boston Red Sox star Ted Williams, who once reportedly said to him, "If I'm in Boston again next year, I'm going to ask Joe Cronin to buy you to get you out of my hair." (Courtesy of Mike Harper.)

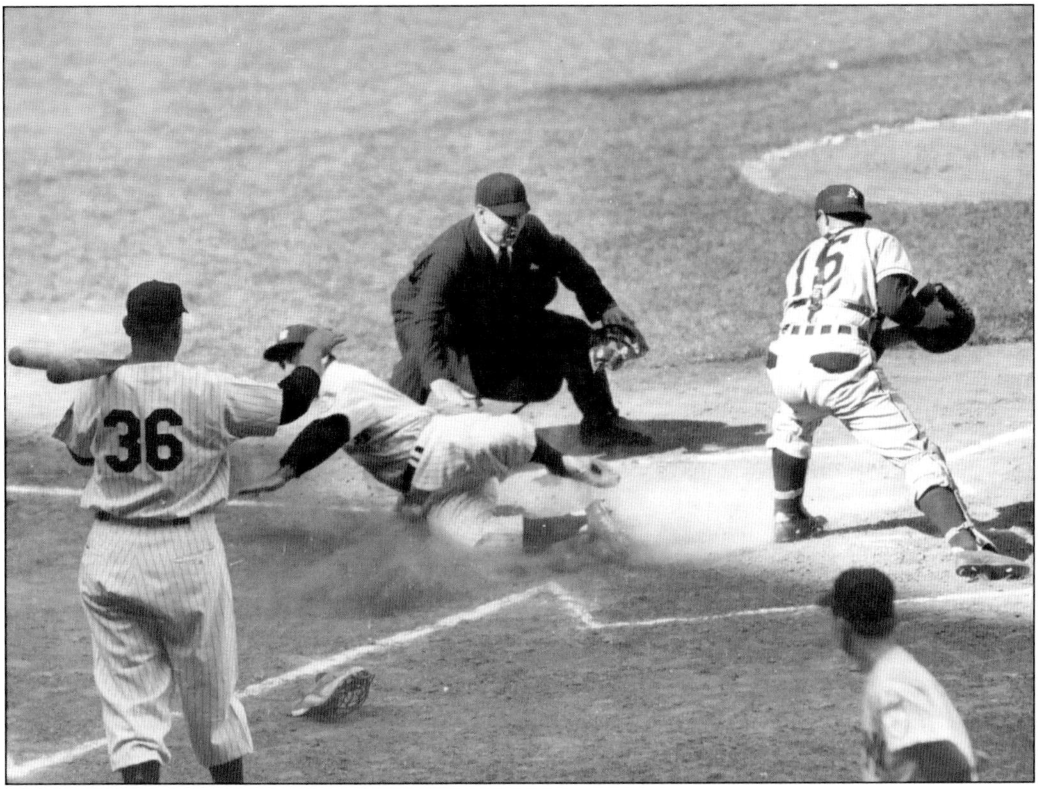

In this 1951 press photograph, as catcher for the Philadelphia Athletics, Joe Tipton takes the throw just a little too late to stop his counterpart, Yankee catcher Yogi Berra, from sliding across home plate. The umpire is Johnny Stevens; the next Yankee batter, at left, is Johnny Mize. (Author's collection.)

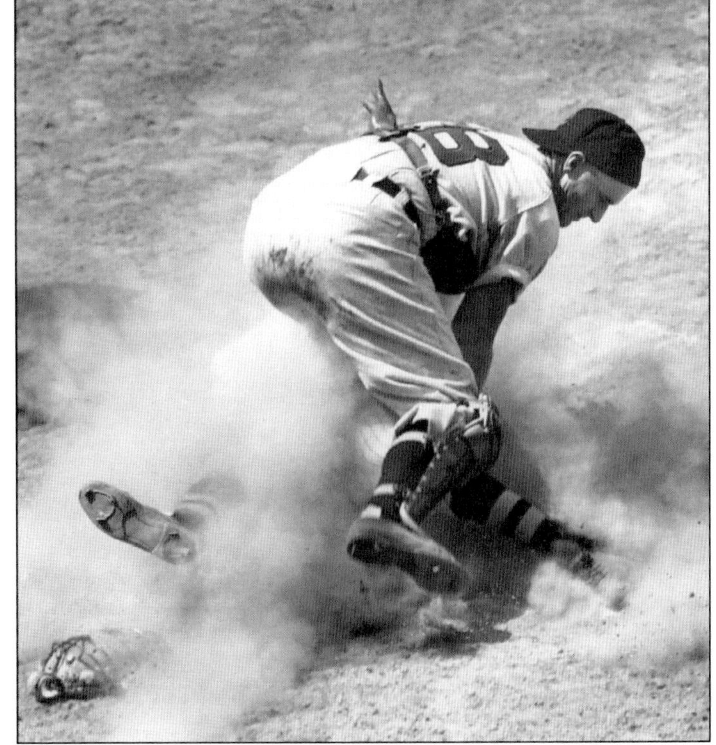

In the June 28, 1952, game between the Cleveland Indians and the Chicago White Sox, Tipton tags White Sox outfielder Minnie Minoso out at home in a close play that resulted in White Sox manager Paul Richards's dismissal from the game. The Indians won, 5-1. (Author's collection.)

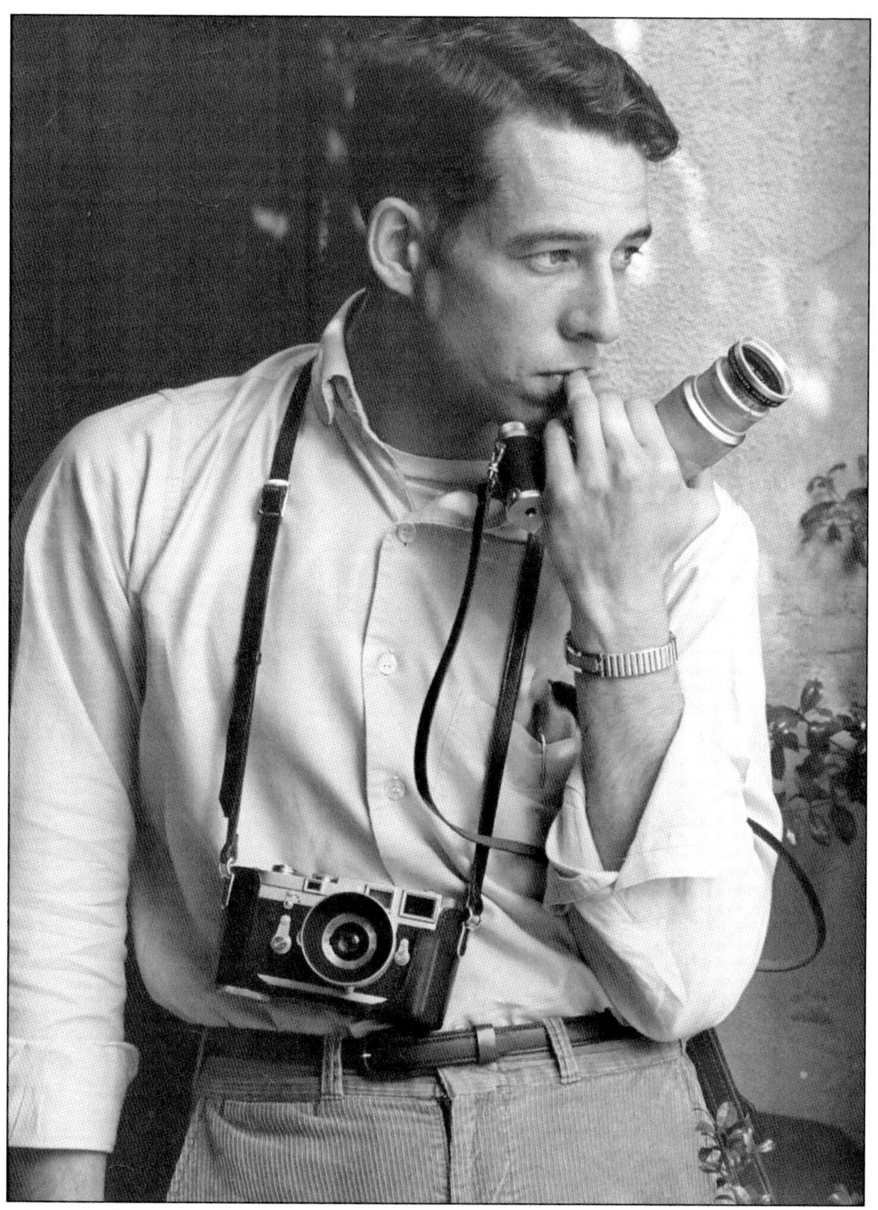

After graduation from Copperhill High School in 1952, class salutatorian Allen Clayton (later known professionally as "Al") joined the Navy as a medical corpsman. In the course of his duties, somebody handed him a camera, and he knew he had found his calling. In 1967, already an established photojournalist, he accepted an assignment from the Southern Regional Council to photograph the conditions of the poor and dispossessed in the American South. The resulting images were used by senators Robert Kennedy and Joseph Clark in their hearings on hunger and malnutrition and are credited with helping to pass the Food Stamp Act. In 1969, his work on assignment for *Look* magazine during the war in Biafra was awarded the Overseas Press Club Citation for Excellence for the year's best coverage of a foreign news event. Many of his photographs of musical icons of the 1960s and 1970s, including Kris Kristofferson, Bob Dylan, Johnny Cash, and others, are displayed in Nashville's Country Music Hall of Fame. He died in 2014 at the age of 79. (Courtesy of Mary Ann and Jennie Clayton.)

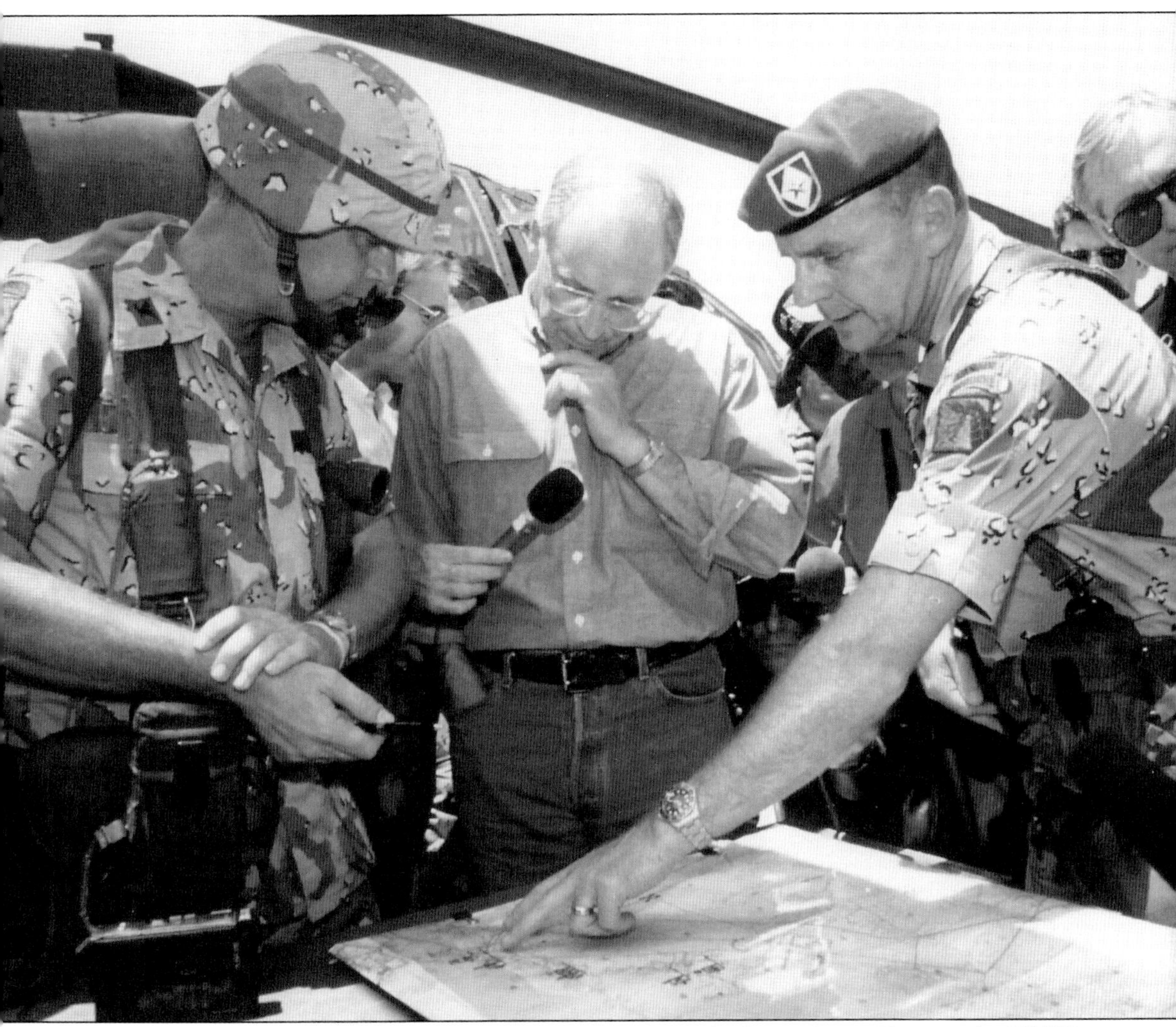

In August 1990, in response to Saddam Hussein's unprovoked attack on Kuwait, the XVIII Airborne Corps became the first paratrooper unit to arrive in Saudi Arabia. Leading the corps of 150,000 troops, 40,000 vehicles, and 1,100 helicopters was Deputy Commanding Gen. Edison E. Scholes, better remembered by his childhood friends in the Basin as Eddie. In this photograph, General Scholes (right) gives a briefing to Secretary of Defense Richard B. Cheney, the unit's first visitor after its arrival in the war zone. In 1989, General Scholes had led the deployment of corps elements to Panama during Operation Just Cause, which resulted in the removal of Manuel Noriega from power. He is also a veteran of two combat tours in Vietnam. Among the awards garnered during his 35-year Army career are two Purple Hearts, a Silver Star and two Bronze Stars for valor, three Bronze Stars for combat service, the Defense Distinguished Service Medal, the Army Commendation Medal for valor, and Vietnamese Gallantry Cross medals with Silver Star and Bronze Star. (Courtesy of Maj. Gen. Edison E. Scholes, USA, Ret.)

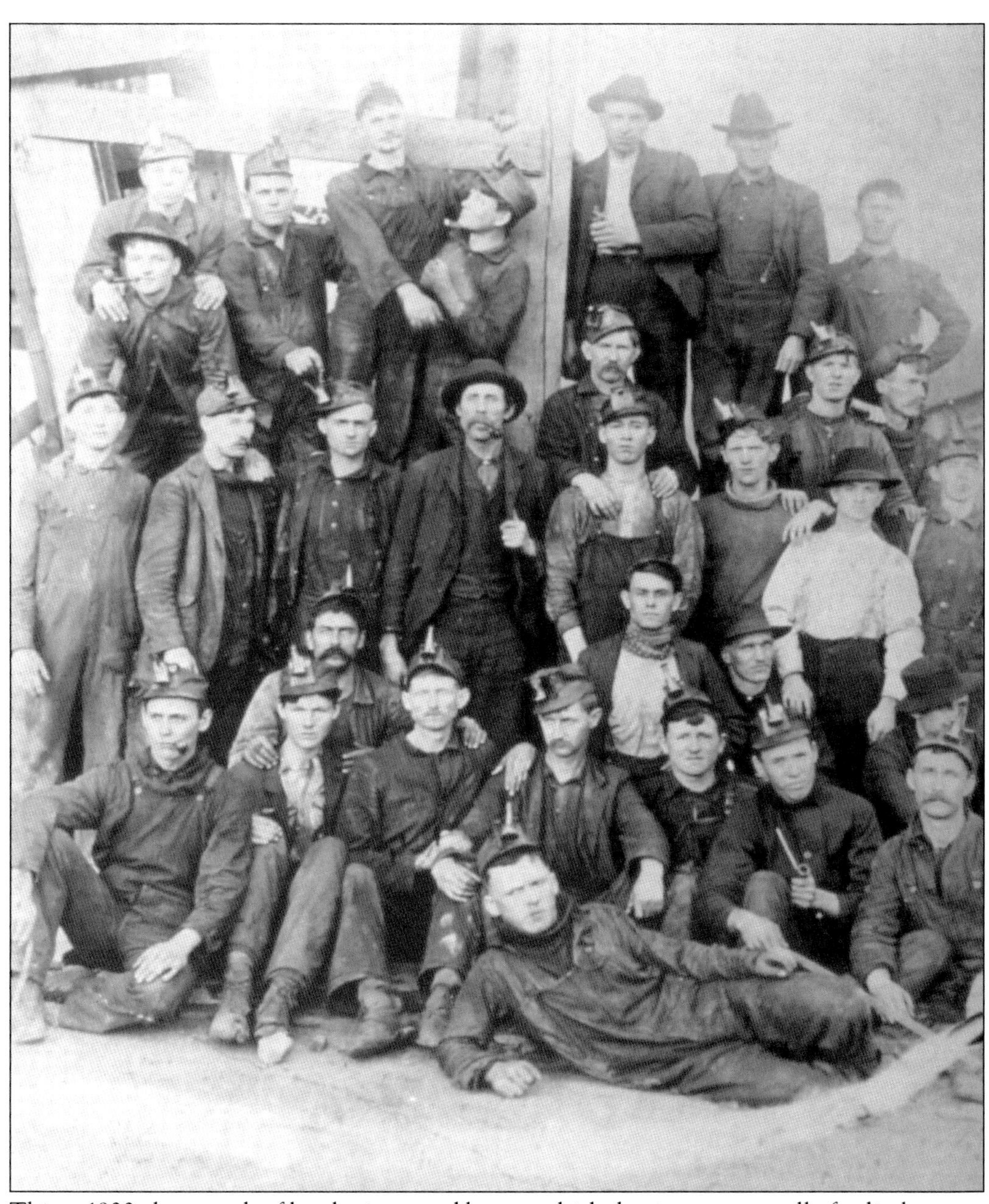

This c. 1900 photograph of local miners and bosses, which dominates one wall of a display area, greets thousands of visitors to the Ducktown Basin Museum every year. Although the faces seem hauntingly familiar, no one has ever been able to identify a relative among them. It seems quite likely, however, that at least some descendants of these early mine workers are still living and working in the Basin. By the time this photograph was taken, which was prior to the advent of carbide headlamps around 1912, both DSC&I and TCC were actively mining in the Ducktown district, which means that the setting could have been any one of about a dozen mines.

"BALLADEER (SERIOUSLY)"

When the last stope has been emptied,
And the drifts are filled with must,
When the skip is parked forever,
And the rails are but streaks of rust,
We shall stand in awe of the silence,
And dig our toes in the crust,
And wait in vain for the whistle
When the ore has turned to dust.

When the buildings are whipt to pieces
By the wind in its lonely flight;
When the grounds are laced with gullies
And the scene is as cheerless as night;
We shall watch the rats from the caverns
Come out and play in the light;
When the ore has all been hoisted
And the rocks and earth unite.

When the crusher no longer rumbles,
And the ladders are frail as thread,
When the pumps have ceased their labors,
And the drills have been put to bed;
We shall wander away in sadness,
O'er trails the miners did tread,
And leave old Burra to slumber
'Neath a sky with blue all spread.

This poem by Copper Basin historian R.E. Barclay Sr., entitled "Ode to Burra Burra," was found in the manuscript of a dramatic narration presented at Copperhill School in 1958 by Polly Morgan and Rev. Allen Cooke. The primary text for the ballad-style narration was written by Virginia Middleton, and Sara Williamson was credited with providing the music. The Barclay poem, which apparently dates from about the same time, is an apparent reflection on the closing of Burra Burra Mine in 1958. Contrary to Barclay's predictions, the mine site is now the location of the Ducktown Basin Museum, and most of its original buildings remain intact. (Author's collection.)

The Ducktown Basin Museum

The Ducktown Basin Museum is located on the site of the historic Burra Burra Mine, which was the headquarters for Tennessee Copper Company and Cities Service mining operations from 1899 through 1975. In 1981, Cities Service donated the site to the museum, which was then in its infancy and was housed in a small storefront location on Main Street in Ducktown. In 1982, the museum formally took over ownership of the property.

The Burra Burra site, now owned by the State of Tennessee and operated as a state historic site under an agreement between the museum and the Tennessee Historical Commission, has been listed in the National Register of Historic Places since 1983. The 16 remaining structures include virtually all of the original mine buildings and outbuildings except for the Burra Burra headframe, which was demolished after the closing of the mine in 1958. The mine office houses the museum's collection and is the only building permanently open to visitors. A self-guided walking tour of the grounds is available during daylight hours, and staff-led tours are available by appointment. A favorite stop for many visitors is an old mine elevator cage, now in use as an open observation platform, overlooking a collapsed and flooded portion of the Burra Burra mine works.

With the closing of the mines in 1987, the museum was able to acquire additional artifacts from the local mining operations. Today's visitors can wander among the giant drills and wrenches, surveyors' instruments, portable telephones, signal code signs, and shift whistles that were part of the miners' everyday lives. Various exhibits detail the history of mining in the Ducktown Basin and the geology of the surrounding area, an audiovisual presentation offers an informative overview, and larger artifacts are displayed on the museum grounds. The gift shop features a variety of merchandise that includes handcrafted jewelry, souvenirs, and books and DVDs of local interest.

The museum is open Monday through Saturday (Tuesday through Saturday in some winter months), with hours varying by season. The most recent information is always available on the museum website at www.ducktownbasinmuseum.com, by telephone at 423-496-5778, or by e-mail at burrahill@ellijay.com. Individual and group membership information and applications are also available on the website.

Bibliography

Barclay, R.E. *The Copper Basin—1890 to 1963*. Knoxville, TN: Cole Printing & Thesis Service, 1975.

———. *Ducktown Back in Raht's Time*. Chapel Hill, NC: The University of North Carolina Press, 1946.

———. *The Railroad Comes to Ducktown*. Cleveland, TN: White Wing Press, 1973.

Science Applications International Corporation. *History of Tennessee Copper Company and Successor Firms at the Copperhill Plant and the Ducktown Mining District, Copper Basin, Tennessee*. Prepared for the US Army Corps of Engineers, Nashville District, and the US Environmental Protection Agency, Region 4, 2008.

Consistent with our mission to preserve history on a local level, this book was printed in South Carolina on American-made paper and manufactured entirely in the United States. Products carrying the accredited Forest Stewardship Council (FSC) label are printed on 100 percent FSC-certified paper.

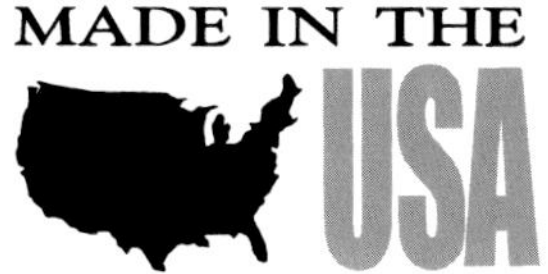